Lecture Notes in Mathematics

Edited by A. Dold, Heidelberg an[d] … Eckmann

384

Functional Analysis and Applications

Proceedings of the Symposium of Analysis
Universidade Federal de Pernambuco
Recife, Pernambuco, Brasil, July 9–29, 1972

Edited by Leopoldo Nachbin
Universidade Federal do Rio de Janeiro and University of Rochester

Springer-Verlag
Berlin · Heidelberg · New York 1974

Prof. L. Nachbin
Avenida Vieira Souto 144, apto. 101
20000 Rio de Janeiro, GB, ZC -95
Brasil

AMS Subject Classifications (1970): 46-02

ISBN 3-540-06752-3 Springer-Verlag Berlin · Heidelberg · New York
ISBN 0-387-06752-3 Springer-Verlag New York · Heidelberg · Berlin

Offsetdruck: Julius Beltz, Hemsbach/Bergstr.

F O R E W O R D

On July 9-29, 1972, a Symposium of Analysis was held at the "Instituto de Matemática da Universidade Federal de Pernambuco", in Recife, Pernambuco, Brazil. The meeting received support from the "Conselho Nacional de Pesquisas (CNPq)", "Coordenação do Aperfeiçoamento do Pessoal de Nível Superior (CAPES)", "Fundação de Amparo à Pesquisa do Estado de S.Paulo (FAPESP)", "Universidade Federal de Pernambuco" and Organization of American States. The Symposium was sponsored by the "Sociedade Brasileira de Matemática".

The organizing committee for this meeting consisted of José Barros Neto, Chaim Samuel Hönig (chairman) and Leopoldo Nachbin (editor). The local organizing committee was formed by Fernando Cardoso, Ruy Luís Gomes and Roberto Ramalho (chairman).

There were over eighty participants from the following countries: Brazil, Colombia, France, Germany, India, Mali, Mexico, Peru, Portugal, Sweden, USA and Venezuela.

The present volume contains the texts of the invited lectures delivered at the Symposium which were submitted to the editor, with the following noteworthy exception. A series of lectures on pseudodifferential and Fourier integral operators offered by François Trèves and expanded into a monograph, was published separately by the "Instituto de Matemática da Universidade Federal de Pernambuco".

Leopoldo Nachbin

CONTENTS

PROLONGEMENT ANALYTIQUE EN DIMENSION INFINIE

par G. Coeuré

1. FONCTIONS ANALYTIQUES

DÉFINITION 1.1. Une fonction f, à valeurs complexes, définie sur un ouvert ω d'un espace de Banach complexe est *analytique* si:

- La restriction de f à toute droite complexe (c.à.d. à un sous-espace affine de dim = 1) est holomorphe.

- f est continue sur ω.

Lorsque $\dim E = +\infty$, f est en général non borné sur les boules fermées contenues dans ω. Ainsi la fonction définie sur un espace de Hilbert séparable par $f(x) = \sum n x_n^n$, où (x_n) est la suite des composantes de x sur une base hilbertienne de E, est analytique sur E et n'est bornée que sur les boules de rayon <1.

La fonction f admet un développement taylorien convergeant normalement localement. Plus précisément:

DÉFINITION 1.2. Un *polynome homogène de degré* n *sur* E est la restriction à la diagonale de $\underbrace{E \times \ldots \times E}_{n \text{ fois}}$, d'une forme n-linéaire sur $E \times \ldots \times E$.

PROPOSITION 1.1. *Soit* f *analytique sur* ω; *pour tout entier* n *et tout* x *de* ω, *il existe un polynome homogène de* $d^o = n$, *continu, noté* $d_x^n f$, *tel que*

$$f(x + h) = \sum_{n=0}^{n=\infty} d_x^n f(h). \tag{1}$$

La série converge normalement en h, *lorsque* h *décrit tout compact de tout ouvert équilibré centré en* x *et contenu dans* ω, *vers* $f(x + h)$.

(2) $$d_x^n f(h) = \int_{|t|=\rho} f(x+th) \frac{dt}{t^{n+1}}, \quad \forall \|h\| < \frac{1}{\rho} d_h(x, \complement\omega)$$

où t *est complexe,* d_h *est la distance dans la direction définie par* h.

$\forall h \in E$, *la fonction* $x \longmapsto d_x^n f(h)$ *est analytique sur* ω.

Les démonstrations des résultats contenus dans cette proposition sont classiques. Le lecteur pourra consulter les ouvrages de HILLE-PHILLIPS [11], M.HERVÉ [10] et la thèse de Ph.NOVERRAZ [14].

Dans la suite nous emploierons les deux notations suivantes:

Si A est une partie d'un espace métrique, A_s désignera l'ouvert obtenu en recouvrant A par des boules de rayon s et centrées sur A.

Pour toute fonction f, définie sur A, à valeurs complexes, on pose $\|f\|_A = \sup_{x\in A} |f(x)|$.

INÉGALITÉS DE CAUCHY

L'inégalité suivante est une conséquence facile de (2). Soit f une fonction analytique sur ω; pour toute partie A de ω et tout $s > 0$ tel que A_s soit contenu dans ω, on a:

(3) $$\|d_x^n f(h)\|_{x\in A} \leq \left\|\frac{h}{s}\right\|^n \|f\|_{A_s}, \quad \forall h \in E.$$

En dimension finie, les démonstrations des propriétés de convexité holomorphe utilisent le fait que l'espace des fonctions, analytiques sur un ouvert ω de $\mathbb{C}^n$, muni de la convergence compacte, est un espace de Frechet.

Si dim $E = \infty$, l'espace fonctionnel ci-dessus est encore complet mais sa topologie n'est évidemment plus définie par une famille dénombrable de semi-normes. On est, ainsi, amené à se servir de topologies plus raffinées.

On ne développe ici que ce qui nous est indispensable pour la suite. Le lecteur trouvera une étude détaillée dans l'ouvrage de L. NACHBIN [13].

L'indice c signifiera que l'espace fonctionnel utilisé est muni de la convergence compacte; l'indice b, de la topologie bornologique associée à la précédente.

$C(\cdot)$ est l'ensemble des fonctions continues sur $(\cdot)$.

Soit X um espace métrique.

DÉFINITION 1.3. Un recouvrement $\mathcal{U}$ de X sera dit *admissible*, s'il est constitué d'ouverts ω tel que ω_s soit contenu dans un autre élément de $\mathcal{U}$, pour s assez petit.

PROPOSITION 1.2. *Pour tout recouvrement ouvert* $\mathcal{V}$ *de* X *et tout* f *de* C(X), *il existe* $\mathcal{U}$ *admissible, plus fin que* $\mathcal{V}$, *tel que* f *soit borné sur les éléments de* $\mathcal{U}$. $\mathcal{U}$ *peut être choisi dénombrable si* $\mathcal{V}$ *l'est.*

DÉMONSTRATION. A ω de $\mathcal{V}$, on associe

$$\omega_n = \{x \in \omega \mid |f(x)| < n\}$$

et

$$\omega_{n,m} = \{x \in \omega_n \mid d(x, \complement\, \omega_n) > \frac{1}{m}\}.$$

Les ouverts $(\omega_{n,m})$ forment le recouvrement $\mathcal{U}$ cherché.

$\mathcal{R}$ désignera l'ensemble des recouvrements admissibles dénombrables de X.

Soit A un sous-espace fermé de $C(X)_c$; pour tout $\mathcal{U}$ de $\mathcal{R}$, on note $A_{\mathcal{U}}$ les sous-espaces de A formé par les f de A tels que

$$\|f\|_\omega < +\infty, \quad \forall\, \omega \in \mathcal{U}.$$

Une fois muni de la convergence uniforme sur les ω de $\mathcal{U}$, $A_{\mathcal{U}}$ est un espace de Fréchet.

PROPOSITION 1.3. $A_b = \varinjlim_{\mathcal{U}\in\mathcal{R}} A_{\mathcal{U}}$; *en particulier* A_b *est tonnelé.*

DÉMONSTRATION. A est la réunion des $A_{\mathcal{U}}$ (Prop. 1.2) et $\varinjlim A_{\mathcal{U}}$ a une topologie plus fine que cette de A_c ; il suffit donc de vérifier que tout borné $\mathcal{B}$ de A_c l'est dans l'un des $A_{\mathcal{U}}$.

Or si $\omega_n = \{x \in X \mid |f(x)| < n \;\; \forall f \in \mathcal{B}\}$, X est la réunion des $\overset{\circ}{\omega}_n$ car sinon il existerait une suite f_n dans $\mathcal{B}$ et une suite convergente (x_n) dans X tels que $|f_n(x_n)| > n$. La suite f_n ne serait donc pas bornée dans A_c. Si $\mathcal{U}$ est le recouvrement construit à partir du recouvrement $(\overset{\circ}{\omega}_n)$ à l'aide de la Prop. 1.2, $\mathcal{B}$ est borné dans $A_{\mathcal{U}}$.

DÉFINITION 1.4. L. Nachbin introduit la topologie suivante:

Une semi-norme p sur A est dite *portée par un compact* K de X si pour tout $\varepsilon > 0$, il existe C_ε tel que

$$p(f) \leqslant C \;\; \|f\|_{K_\varepsilon} \quad \forall f \in A.$$

La topologie A_N est celle définie par ce type de semi-normes.

PROPOSITION 1.4. *On a* $A_b \longmapsto A_N \longmapsto A_c$.

DÉMONSTRATION. Le morphisme $A_N \longmapsto A_c$ est trivial; il suffit donc là aussi de vérifier que tout borné $\mathcal{B}$ de A_c l'est dans A_N; à celle fin étant donnée une semi-norme portée par le compact K, on peut recouvrir K par un nombre fini d'ouverts appartenant au recouvrement

$\mathcal{U}$ construit à la Prop. 1.3 et pour lequel $\mathcal{B}$ est borné dans $A_{\mathcal{U}}$. Dans ces conditions $\|\mathcal{B}\|_{K_\varepsilon}$ est fini pour ε assez petit.

2. CONVEXITÉ DES ENVELOPPES D'HOLOMORPHIE

X est une variété connexe étalée au dessus de E au moyen d'un homéomorphisme local p.

d_X désignera la fonction distance sur X définie comme le rayon de la plus grande boule de E centrée en p(x) et homéomorphe par p à un voisinage de x. Pour toute partie T de X, on pose

$$d(T) = \inf_{x \in T} d(x).$$

Une fonction f, continue sur X, à valeurs complexes, est analytique si pour tout x de X, il existe un voisinage ω de p(x) et une fonction g analytique dans ω telle que $f = g \circ p$ dans ω.

$\mathcal{O}(X)$ désignera l'ensemble des fonctions analytiques sur X.

Étant donnée une autre variété X' étalée sur E, un morphisme de X vers X' est une application continue de X dans X' rendant commutatif le diagramme suivant:

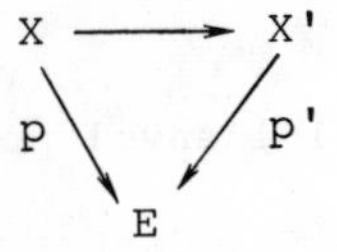

Étant donnée une partie $\mathcal{A}$ de $\mathcal{O}(X)$, un morphisme $j: X \to X'$ est une extension analytique de $\mathcal{A}$, s'il existe une partie $\mathcal{A}'$ de $\mathcal{O}(X')$ tel que le morphisme transposé j^* de $\mathcal{O}(X')$ dans $\mathcal{O}(X)$ transforme $\mathcal{A}'$ en $\mathcal{A}$.

$\tilde{X}$ est dite *l'enveloppe d'holomorphie* de $\mathcal{A}$, si $\tilde{X}$ est une extension analytique de $\mathcal{A}$, maximale dans le sens suivant:

pour toute autre extension analytique X' de $\mathcal{A}$, il existe un morphisme de X' vers $\tilde{X}$ rendant commutatif le diagramme:

$$\begin{array}{ccc} X & \longrightarrow & \tilde{X} \\ & \searrow \quad \nearrow & \\ & X' & \end{array}$$

Toute famille $\mathcal{A}$ dans $\mathcal{O}(X)$ admet un enveloppe d'holomorphie que l'on notera $\mathcal{E}(\mathcal{A})$. Nous ne feron pas sa construction ici. Elle est décrite complétement par SCHOTTENLOHER dans [15].

Les problémes suivants se posent naturellement sur les enveloppes puisqu'ils admettent une réponse positive lorsque $\dim E < \infty$. Nous donnons les résultats les plus importants au sujet des trois premiers et développons le quatrième. Introduisons d'abord la notion suivante:

DÉFINITION 2.1. X est dite *$\mathcal{A}$-convexe* si pour tout compact K de X, on a $d(\hat{K}_{\mathcal{A}}) > 0$, où $\hat{K}_{\mathcal{A}}$ est l'enveloppe de K relativement à $\mathcal{A}$ et est défini par:

$$\hat{K}_{\mathcal{A}} = \{x \in X \mid \|f(x)\| \leqslant \|f\|_K \quad \forall\, f \in \mathcal{A}\}.$$

X est dite *holomorphiquement convexe* si pour tout K compact $\hat{K}_{\mathcal{O}(x)}$ est compact.

PROBLÈME 1. $\mathcal{E}(\mathcal{A})$ est-il l'enveloppe d'holomorphie d'une famille réduite à une seule fonction?

Un exemple remarquable dû à HIRSCHOWITZ [12], montre que si E n'est pas séparable, la réponse peut être négative. Toute fois le théorème important suivant de GRUMAN et KISELMAN [8] donne une réponse positive au problème lorsque $\mathcal{E}(\mathcal{A})$ est univalente et E est un espace à base dénombrable.

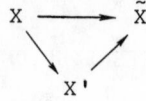

THÉORÈME 2.1. *Étant donné un domaine pseudo-convexe X d'un espace de Banach à base dénombrable (X est coupé par tout sous-espace affine de dimension finie suivant un ouvert pseudo-convexe), il existe une suite (x_n) dans X, adhérente à tout point frontière de X, et f dans $\mathcal{O}(X)$ tels que $|f(x_n)| \geqslant n$ pour tout n.*

Sous réserve que les hypothèses utilisées par le théorème précedent soient réalisées, le problème suivant a en conséquence une réponse positive.

PROBLÈME 2. Si X est holomorphiquement convexe, X est-il l'enveloppe d'holomorphie de $\mathcal{O}(X)$?

Considérons maintenant le problème réciproque au problème 2.

PROBLÈME 3. $E(A)$ est-il $\tilde{A}$-convexe où $\tilde{A}$ est la famille obtenue à partir de A par prolongement analytique à $E(A)$?

Elargissons la notion de A-convexité de la façon suivante:

DÉFINITION 2.2. Une partie T de X sera dite *A-bornée* si

$$\|f\|_T < \infty, \quad \forall f \in A.$$

L'introduction de cette notion est motivée par la découverte par DINEEN [5] d'espaces E (ℓ_∞ par exemple) dans lesquels on peut construire les parties $\mathcal{O}(E)$-bornées et non relativement compactes. Toutefois si E est séparable seules les parties relativement compactes sont $\mathcal{O}(E)$-bornées (HIRSCHOWITZ [13], SCHOTTENLOHER [15]).

$E(A)$ étant l'enveloppe d'holomorphie de $\tilde{A}$, on peut supposer dans l'étude du problème 3 que X est l'enveloppe d'holomorphie de A. Soit T une partie de X, $d(T) > 0$, et posons $\mu_A(T)$ = la borne supèrieure des $r \in]0, d(T)[$ tels que pour tout a de E ($\|a\| = 1$), le recouvrement de T par des cercles de rayon r, parallèles à a, centrés sur

T, soit une partie A-bornée, si de tels r existent; dans le contraire on pose $\mu_A(T) = 0$.

PROPOSITION 2.2. Si X est l'enveloppe d'holomorphie d'un sous-espace A du type A_U de $\mathcal{O}(X)$, stable par dérivation, alors pour toute partie T telle que $d(T) > 0$ on a:

$$d(\hat{T}_A) \geqslant \mu_A(T).$$

Remarquons que cette dernière inégalité ne donne une information sur l'enveloppe $\hat{T}_A$ que si $\mu_A(T) > 0$; ce qui implique que T est A-bornée.

Le problème de savoir si les parties A-bornées verifint $\mu_A(T) > 0$ semble ne pas avoir été étudié.

Lorsque T est compact on a $\mu_A(T) = d(T)$ et on obtient le corollaire suivant qui donne une réponse positive au problème 3.

COROLLAIRE 2.3. *Si* X *est l'enveloppe d'holomorphie d'un sous-espace* A *du type* A_U *de* $\mathcal{O}(X)$, *stable par dérivation, alors* X *est* A-*convexe et pour tout compact* K *de* X, *on a:*

$$d(\hat{K}_A) = d(K).$$

Revenons à la proposition 2.2 et donnons l'idée directrice de la démonstration qui est détaillée dans [2].

Étant donné x dans $\hat{T}_A$ et b dans E ($\|b\| < \mu_A(T)$), on montre que la série $\sum d_x^n f(b+\alpha)$ définit une fonction analytique $h_b(\alpha)$ pour $\|\alpha\|$ assez petit. A cette fin on montre que le recouvrement de T par des cercles de rayon $\|b\|$, parallèles à b et centrés sur T, est contenu dans un ouvert où f est bornée; on applique alors les inégalités de Cauchy. On constate enfin que les germes h_b (pour $\alpha = 0$) forment un voisinage du germe $(f \circ p_x^{-1})_{p(x)}$ homéomorphe à la boule de E de rayon $\mu_A(T)$.

PROBLÈME 4. Si X est A-convexe, X est'il holomorphiquement convexe?

Si X est un domaine de E, ou plus généralement si X a un nombre fini de feuillets, la réponse est positive et sa démonstration est identique à celle utilisée lorsque dim E < ∞ (cf. par exemple GUNNING et ROSSI [9]).

3. UN RÉSULTAT D'APPROXIMATION

ω est un ouvert de $\mathbb{C}^n$, ϕ est une fonction de classe C^2 sur ω; $L^2_q(\phi)$ désigne l'espace de Hilbert formé par les formes différentielles en $d\bar{z}$ de $d^o = q$, à coefficients de carré intégrable par rapport à la mesure $e^{-\phi}\cdot dz\cdot d\bar{z}$.

Nous avons besoin du résultat suivant que est conséquence de la théorie d'Hörmander sur l'opérateur $\bar{d}$.

THÉORÈME H. *Soit* ω *un ouvert d'holomorphie, borné de* $\mathbb{C}^n$. *Il existe deux fonctions* ϕ_1 *et* ϕ_2 *de classe* C^2 *sur* ω *telles que l'équation* $\sum_{i=1}^{r} \alpha_i \bar{d} u_i = v$ *admette une solution* $u = (u_i)$ *unique dans* dom $T \cap \overline{\mathrm{Im}\, T^*}$ *pour tout* v *de* $L^2_1(\phi_2)$ *vérifiant* $\bar{d}v = 0$ *et*

$$\int \|v\|^2 (\sum |\alpha_i|^2)^{-1} e^{-\phi_2} dz\cdot d\bar{z} < \infty$$

et la solution u *vérifie l'inégalité suivante*

$$(1) \qquad \int \|u\|^2 e^{-\phi_1} dz\cdot d\bar{z} \leq \int \|v\|^2 (\sum |\alpha_i|^2) e^{-\phi_2} dz\cdot d\bar{z} < \infty.$$

$\alpha = (\alpha_i)$ *est donné dans* $[\mathcal{O}(\bar{\omega})]^r$, T *est l'opérateur hilbertien de* $[L^2_o(\phi_1)]^r$ *vers* $L^2_1(\phi_2)$ *défini par* $T(u) = \sum \alpha_i \bar{d} u_i$.

De plus une fonction h *de* $C(\omega)$ *étant donnée,* ϕ_2 *peut être choisi pour que* h *soit dans* $L^2(\phi_2)$.

Dans la suite E est un espace de Banach.

COROLLAIRE 3.1. *Pour toute forme* $v(z,a)$ *de* $d^o = 1$ *en* $d\bar{z}$, *dont les coefficients de* $(\sum|\alpha_i|^2)^{-1/2}\cdot v$ *sont continus sur* $\bar{\omega}\times E$ *et polynomiaux homogènes de* $d^o = k$ *en* $a \in E$, *l'équation* $T(u) = v$ *admet une solution* $u(z,a)$ *polynomiale homogène de* $d^o = k$ *en* a, *appartenant à* $L^2_o(\phi_1)$ *pour* a *fixé, vérifiant* (1).

DÉMONSTRATION. On choisit ϕ_2 pour que les coefficients de $(\sum|\alpha_i|^2)^{1/2}\cdot v$ soient dans $L^2(\phi_2)$; puis on résoud l'équation $T(u\ ) = v^{\#}$, à l'aide du théorème H, dans laquelle $v^{\#}$ est la forme k-linéaire symétrique sur E associée à v. L'unicité de la solution impose la linéairité de $u^{\#}$. La solution cherchée est le polynome associé à $u^{\#}$.

THÉORÈME 3.2. *Soit* K *un compact holomorphiquement convexe dans* ω *et* U *un voisinage de* K. *Pour toute fonction* p *analytique sur* $U\times E$ *polynomiale homogène de* $d^o = k$ *sur* E, *pour tout* $\varepsilon > 0$, *pour tout borné* B *de* E, *il existe* $\tilde{p}$, *analytique sur* $\omega\times E$, *polynomiale homogène de* $d^o = k$ *sur* E, *tel que*

$$\|\tilde{p} - p\|_{K\times B} < \varepsilon.$$

DÉMONSTRATION. Il existe un polyèdre analytique W tel que $K\subset W\subset\overline{W}\subset U$, W est défini par $\{z\in U \mid |g_i(z)| < 1,\ \forall\, i\}$ avec $g = (g_i) \in [\mathcal{O}(\omega)]^r$. Soient Δ_1 et Δ_2 les polydisques unités relativement aux variables z et ζ de ω et $\mathbf{C}^r$; on choisit M pour que $\phi: z \longmapsto (z/M, g(z))$ applique W dans $\Delta_1\times\Delta_2$. Il existe $\rho < 1$ tel que $\phi(K)$ soit contenu dans $\Delta_1\times\rho\cdot\Delta_2$ et $Q = \phi(W) \cap (\Delta_1\times\overline{\rho\cdot\Delta_2})$ est compact dans $\Delta_1\times\Delta_2$; il existe donc ψ dans $C^\infty(\Delta_1\times\Delta_2)$, à support compact dans $W/M\times\Delta_2$, valant 1 dans un voisinage de Q.

On peut appliquer le Corollaire 3.1 à l'équation

$$T(u) = -\bar{d}\psi\cdot(p\circ\phi^{-1})$$

en prenant les fonctions $\zeta_i - g_i\circ\phi^{-1}$ comme fonction α_i. Si $u=(u_i)$ est la solution de cette équation, la fonction G définie par

$$\psi\cdot(p\circ\phi^{-1}) + \sum\alpha_i u_i$$

est analytique sur $\Delta_1\times\rho\cdot\Delta_2$ et polynomiale en a.

Montrons que G est continue sur $\Delta_1\times\rho\cdot\Delta_2\times E$. A cette fin appliquons l'inégalité de la moyenne à G^2 sur une boule centrée en (z,ζ), fermée, et contenue dans $\Delta_1\times\rho\cdot\Delta_2$. On obtient à l'aide de (1):

$$|G(z,\zeta,a)|^2 \leqslant k_1\|\psi\cdot(p\circ\phi^{-1})\|^2_{L^2_0(\phi_1)} + k_2\|\bar{d}\psi(p\circ\phi^{-1})\|^2_{L^2_1(\phi_2)}.$$

Or p est continu sur $\overline{W}\times E$ et ψ s'annule hors de $W\times E$, on en déduit que les normes du second membre sont bornées lorsque a décrit un borné de E; G est donc continu de la variable a.

L'extension du théorème de Hartogs aux espaces de Frechet démontrée par NOVERRAZ dans [14] permet de conclure à l'analycité de G sur $\Delta_1\times\rho\cdot\Delta_2\times E$.

Le polynome $\tilde{p}$ s'obtient en composant avec ϕ la somme des N premiers termes du développement taylorien de G à l'origine, pour N assez grand. On remarque en fin que $G\circ\phi = p$ sur $K\times E$.

DÉFINITION 3.1. Soit X un ensemble; une partie Ω de $X\times E$ sera dite X-*équilibrée* si Ω se projette <u>sur</u> X par l'application canonique de $X\times E$ dans X et si:

$$(x,a)\in\Omega \Longrightarrow (x,ta)\in\Omega \quad \forall\ |t|\leqslant 1.$$

Les deux théorèmes qui suivent sont des applications du théorème précédent. Comme il n'en sera pas fait usage dans la suite nous les donnons sans démonstration. Ces démonstrations sont écrites dans [2].

THÉORÈME 3.3. *Soit* Ω *un ouvert* K-*équilibré, où* K *est un compact holomorphiquement convexe dans un ouvert d'holomorphie* ω *de* $\mathbf{C}^n$. *Soit* f *une fonction analytique dans un voisinage de* Ω.

Alors pour tout $\varepsilon > 0$, *pour tout compact* Q *de* Ω, *il existe* p *analytique sur* $\omega \times E$, *polynomiale sur* E, *tel que:*

$$\|f - p\|_Q < \varepsilon .$$

DÉFINITION 3.2. Soit A une partie d'un espace de Banach. $\mathcal{O}^*(A)$ désignera l'ahérence dans $C(A)_c$ des fonctions analytiques au voisinage de A.

THÉORÈME 3.4. *Soit* X *un ouvert d'holomorphie de* $\mathbf{C}^n$ *et* Ω *un ouvert* X-*équilibré de* $X \times E$.

L'espace $\mathcal{O}(\Omega)_N$ *est la limite projective des espaces*

$$\mathcal{O}^*[(K \times E) \cap \Omega]_N$$

lorsque K *décrit les compact de* X.

4. PROLONGEMENT ANALYTIQUE SUR LES OUVERTS X-ÉQUILIBRÉS

Cette étude prolonge celle de S.DINEEN [5] sur les ouverts de Runge. Si X est réduit à un point x, les ouverts $\{x\}$-équilibrés ne sont qu'un cas particulier des ouverts de Runge, mais en prenant pour X un ouvert holomorphiquement convexe d'un espace de Banach F, on obtiendra des résultats sur des ouverts plus généraux.

On suppose d'abord seulement que X est un espace métrique. Soit Ω un ouvert X-équilibré de $X \times E$; Ω_x désignera la section de Ω à x constant où x est le point générique de X; f_x est la trace sur Ω_x d'une fonction définie sur Ω.

Un f de $C(\Omega)$ sera dite *analytique* en E si pour tout x de X, f_x est analytique sur Ω_x.

Soit A un sous-espace fermé de $C(\Omega)_c$, formé de fonctions analytiques en E et tels que $(x, a) \longmapsto d_0^n f_x(a)$ soit dans A pour tout f dans A, et tout entier n.

PROPOSITION 4.1. *La série* $\sum_{n=0}^{n=\infty} d_0^n f_x(a)$ *converge vers* f *dans* A_b.

DÉMONSTRATION. Soit ε_n une suite de réels avec $\varepsilon_n \longrightarrow 1$ et $\varepsilon_n^n \longrightarrow +\infty$.

Étant donné un compact Q de Ω dont $\tilde{Q}$ est l'enveloppe X-équilibré, on vérifie à l'aide de la Proposition 1.1 que

$$\| \varepsilon_n^n [f(x, a) - \sum_0^n d_0^k f(x)(a)] \|_Q \leqslant \left(\frac{\varepsilon_n}{\rho}\right)^n \|f\|_{\rho \cdot \tilde{Q}}$$

avec $\rho > 1$ et $\rho \cdot \tilde{Q} \subset \Omega$.

La suite

$$g_n = \varepsilon_n^n [f(x, a) - \sum_0^n d_0^k f(x)(a)]$$

est donc bornée dans A_c et A_b. Si p est une semi-norme continue sur A_b, $\frac{1}{\varepsilon_n^n} p(g_n)$ tend vers 0.

REMARQUE. Les hypothèses imposées à A sont vérifiées dans de nombreux exemples concrets; ainsi si X est un ouvert de $\mathbb{R}^n$ on peut pendre pour A l'espace des fonctions analytiques en E et harmoniques sur les sections de Ω par les sous-espaces parallèles à X.

Lorsque X est une partie quelconque d'un espace de Banach, $\mathcal{O}^*(\Omega)$ est un $\mathcal{A}$ particulier.

Lorsque X est un espace compact et E est par exemple un espace de Banach convenable, on peut démonstrer que les topologies $\mathcal{A}_N$ et $\mathcal{A}_b$ sont identiques; la démonstration est écrite dans [2]. Nous n'utiliserons pas ce résultat ici.

On suppose maintenant que X est une partie d'un espace de Banach F.

Pour toute partie A de X, on pose $\Omega_A = (A \times E) \cap \Omega$. Soit $\mathcal{F}_A$ la famille des ouverts ω du sous-espace topologique $A \times E$ de $F \times E$, dont les sections ω_x sont connexes et contiennent $(\Omega_A)_x$, pour lesquels tout f de $\mathcal{O}^*(\Omega_A)$ se prolonge en une fonction $\tilde{f}$ de $C(\omega)$ analytique en E et tels que l'application de $\mathcal{O}^*(\Omega_A)_c$ dans $C(\omega)_c$ ainsi obtenue soit continue.

Par Zornification, on constante que $\mathcal{F}_A$ possède des éléments maximaux pour la relation d'inclusion; $\tilde{\Omega}_A$ sera l'un d'entre-eux et $\tilde{f}_A$ sera le prolongement de f à $\tilde{\Omega}_A$.

PROPOSITION 4.2. $\tilde{f}_A$ *est dans* $\mathcal{O}^*(\tilde{\Omega}_A)$ *et les fonctions de* $\mathcal{O}^*(\tilde{\Omega}_A)$ *polynomiales en* E *sont denses dans* $\mathcal{O}^*(\tilde{\Omega}_A)_c$.

DÉMONSTRATION. Étant donné un compact $\tilde{Q}$ de $\tilde{\Omega}_A$, il existe un compact Q de Ω_A pour lequel on a: $\|\tilde{f}_A\|_{\tilde{Q}} \leq \|f\|_Q$, pour tout f de $\mathcal{O}^*(\Omega_A)$. D'autre part, d'après la Proposition 4.1, il existe un voisinage V de Ω_A dans $F \times E$, une fonction g de $\mathcal{O}(V)$ et un entier N tels que:

$$\|f(x, a) - \sum_{n=0}^{n=N} d_0^n g_x(a)\|_Q < \varepsilon.$$

La fonction $\sum_{n=0}^{n=N} d_o^n g_x(a)$ est dans $\mathcal{O}[(V \cap F) \times E]$ et approche $\tilde{f}_A$ sur $\tilde{Q}$ à ε près.

PROPOSITION 4.3. *Si* A *est contenu dans* $\overset{\circ}{B}$ *et si la distance de* A *à la frontière de* $\overset{\circ}{B}$ *est strictement positive,* $\tilde{\Omega}_A$ *est contenu dans* $\tilde{\Omega}_B$.

DÉMONSTRATION. Dans le cas contraire il existerait un point y se projetant en x sur A, qui appartiendrait à la frontière de la composante convexe de $(\tilde{\Omega}_B)_x \cap (\tilde{\Omega}_A)_x$ contenant $(\Omega_A)_x$, dans $(\tilde{\Omega}_A)x$.

Si $\tilde{f}_B$ est analytique an voisinage de $\tilde{\Omega}_B$, $\tilde{f}_B$ et $\tilde{f}_A$ coïncidant en y et il existe un compact Q de Ω_A pour lequel:

$$|\tilde{f}_A(y)| = |\tilde{f}_B(y)| \leqslant \|f\|_Q.$$

Soit $r > 0$, inférieur à la distance de Q à la frontière de $(\overset{\circ}{B} \times E) \cap \Omega_B$. Notons Γ la boule centrée en y et de rayon r dans $F \times E$.

Le développement taylorien de $\tilde{f}_B$ en y converge uniformément sur tout compact K de Γ et on peut associer à K, à l'aide des inégalités de Cauchy, un compact K' de Ω_B tel que:

$$\text{(1)} \qquad \|\text{Tayl. } \tilde{f}_B\|_K \leqslant \|f\|_{K'}.$$

Pour tout h dans Γ, la forme linéaire $\tilde{f} \longmapsto \text{Tayl. } \tilde{f}_B(h)$ est donc continue sur une partie partout dense de $\mathcal{O}^*(\tilde{\Omega}_B)_c$ d'après la proposition 4.1, elle se prolonge donc à $\mathcal{O}^*(\tilde{\Omega}_B)$ sur lequel (1) reste vrai. La fonction Tayl. $\tilde{f}_B$ est donc limite uniforme sur tout compact de Γ d'une suite de fonctions analytiques, elle est analytique.

Soit h dans $\Gamma \cap \tilde{\Omega}_B$; toujours d'après la Proposition 4.1, il existe une suite g_n dans $\mathcal{O}^*(\Omega_B)$, polynomiale en E, tendant uniformément vers $\tilde{f}_B$ sur $K' \cup \{h\}$. Les g_n sont analytiques sur Γ, on a donc:

$$\text{Tayl. } \tilde{f}_B(h) = \lim \text{Tayl. } g_n(h) = \lim g_n(h) = \tilde{f}_B(h) .$$

Toute fonction f de $\mathcal{O}^*(\Omega_B)$ se prolongerait donc à $\tilde{\Omega}_B \cup \Gamma$ et les espaces $\mathcal{O}^*(\tilde{\Omega}_B)_c$ et $\mathcal{O}^*(\Omega_B \ \Gamma)_c$ seraient isomorphes à cause de (1); ce qui est constradictoire avec la maximalité de $\tilde{\Omega}_B$.

PROPOSITION 4.4. *Si* A *est holomorphiquement convexe par rapport à* $\mathcal{O}^*(A)$, *pour tout compact* Q *de* $(\overset{\circ}{A} \times E) \cap \tilde{\Omega}_A$, $\hat{Q}_{\mathcal{O}^*(\tilde{\Omega}_A)}$ *est compact et sa distance parallèlement à* E *au complémentaire de* $\tilde{\Omega}_A$ *est supérieure ou égale à la distance de* Q *au complémentaire de* $(\overset{\circ}{A} \times E) \cap \tilde{\Omega}_A$.

DÉMONSTRATION. $\hat{Q}$ est contenu dans le produit de enveloppe $\mathcal{O}^*(A)$-convexe de la projection de Q sur F par l'enveloppe linéaire fermée de la projection de Q sur E; $\hat{Q}$ est donc contenu dans un compact de $A \times E$; il reste à vérifier que la distance de $\hat{Q}$ à E au complémentaire de $\tilde{\Omega}_A$ est strictement positive.

Le même raisonnement qu'à la Proposition 4.2, montre que pour tout y de $\hat{Q}$, le développement taylorien de tout f de $\mathcal{O}^*(\tilde{\Omega}_A)$ converge dans la boule de centre y et de rayon égal à la distance de Q au complémentaire de $(\overset{\circ}{A} \times E) \cap \tilde{\Omega}_A$, vers une fonction qui prolonge f. Compte tenu de la maximalité de $\tilde{\Omega}_A$, on obtient ainsi la propriété cherchée.

THÉORÈME 4.5. *Si* E *et* F *sont des espaces de Banach à base, si* X *est un domaine holomorphiquement convexe de* F, $\tilde{\Omega}_X$ *est l'enveloppe d'holomorphie de* Ω, *qui par conséquent est univalente.*

DÉMONSTRATION. $\tilde{\Omega}_X$ est holomorphiquement convexe d'après la Proposition 4.4, il suffit d'appliquer le Théorème 2.1 de GRUMAN et KISELMANN [8].

REMARQUE. Si X est un espace compact quelconque, et E un espace de Banach convenable, on peut montrer que l'ouvert $\tilde{\Omega}_X$, construit à partir d'un espace A vérifiant les hypothèses de la Proposition 4.1, vérifie la propriété suivante: Pour toute suite y_n tendant vers un point frontière de $\tilde{\Omega}_X$, il existe des y'_n arbitrairement voisins des y_n dans $\tilde{\Omega}_X$ et un f de A tel que $\sup|\tilde{f}(y'_n)| = +\infty$. La démonstration est faite dans [2] et généralise des résultats antérieurs [6],[4].

Lorsque F est de dimension finie (faute de mieux!) on prouve maintenant la propriété suivante de $\tilde{\Omega}_X$.

THÉORÈME 4.6. *Si* X *est un ouvert d'holomorphie de* F *et* K_m *est une exhaustion de* X *par une suite* K_m *de compacts,* $\tilde{\Omega}_X$ *est la réunion des* $\tilde{\Omega}_{K_m}$.

DÉMONSTRATION. Les $\tilde{\Omega}_{K_m}$ forment une suite croissante, d'après la Proposition 4.3, dont la réunion est un ouvert $\tilde{\Omega}$ de $X \times E$ contenant Ω. D'après le Théorème 2.1, il reste à vérifier que $\tilde{\Omega}$ est holomorphiquement convexe.

Soit K un compact de $\tilde{\Omega}$, la projection de K sur X a une enveloppe $\mathcal{O}(X)$-convexe qui est compact et $\hat{K}_{\mathcal{O}(\tilde{\Omega})}$ est contenu et compact dans $K_m \times E$ pour m assez grand; il reste à verifier que la distance de $\hat{K}_{\mathcal{O}(\tilde{\Omega})}$ à la frontière de $\tilde{\Omega}$ est non nulle et pour cela en appliquant la Proposition 4.4 de prouver que tout point y de $\hat{K}_{\mathcal{O}(\tilde{\Omega})}$ est contenu dans $\hat{K}_{\mathcal{O}^*(\tilde{\Omega}_{K_m})}$ pour m assez grand.

Choisissons m assez grand pour que $y \in \tilde{\Omega}_{K_m}$ et que l'enveloppe $\mathcal{O}(X)$-convexe de $K \cup \{y\}$ soit contenu dans $\overset{\circ}{K}_m \times E$. Dans ces conditions, par application du théorème 3, pour tout $\varepsilon > 0$ et toute fonction p de $\mathcal{O}^*(\tilde{\Omega}_{K_m})$ polynomiale en E il existe une fonction p_ε de $\mathcal{O}(\tilde{\Omega})$ polynomiale en E tel que

$$\|p - p_\varepsilon\|_{K\cup\{y\}} < \varepsilon.$$

Il en résulte:

$$|p(y)| \leqslant |p_\varepsilon(y)| + \varepsilon \leqslant \|p_\varepsilon\|_K + \varepsilon < \|p\|_K + 2\varepsilon.$$

Par passage à la limite on obtient: $|p(y)| \leqslant \|p\|_K$.

Cette inégalité s'étend aux fonctions de $\mathcal{O}^*(\tilde{\Omega}_{K_m})$, puisque les fonctions polynomiales en E sont denses dans $\mathcal{O}^*(\tilde{\Omega}_{K_m})_c$ (Proposition 4.2).

Université de Nancy
Département de Mathematiques
Nancy
FRANCE

BIBLIOGRAPHIE

[1] E.BISHOP, *Holomorphic completion, analytic continuation, and the interpolation of semi-norms*, Ann.Math. 78(1963)

[2] G.COEURÉ, *Analytic functions on manifold spread over Banach spaces* (à paraître)

[3] *Fonctions analytiques et plurisousharmoniques*, Ann. Inst. Fourier 20 (1970)

[4] *Fonctionnelles analytiques sur certains espaces de Banach*, Ann. Inst. Fourier 21 (1971)

[5] S.DINEEN, *Bounding subsets of a Banach space*, Math. Ann. 192 (1971)

[6] *Runge domains in Banach spaces*, Proc. Roy. Irish Ac. 71 (1971)

[7] *Holomorphic functions on C_o-modules*, Math. Ann. 196 (1972)

[8] L.GRUMAN - C.O.KISELMAN, *Le problème de Lévi dans les espaces de Banach à base*, C.R.Ac.Sc. Paris (1972)

[9] GUNNING-ROSSI, Analytic functions of sev. complex. var., Prentice Hall (1965)

[10] M.HERVÉ, Analytic and plurisubharmonic functions, Springer Lectures Notes 198 (1971)

[11] HILLE-PHILIPS, *Functional analysis and semi-groups*, Amer. Math. Soc. 31 (1957)

[12] A.HIRSCHOWITZ, *Sur le prolongement des variétés analytiques réelles*, C.R.Ac.Sc. Paris (1969)

Bornologie des espaces de fonctions analytiques en dimension infinie, Sém. P.Lelong, Paris (1969)

[13] L.NACHBIN, Topologies on spaces of holomorphic mappings, Erg. d. Math. 47 (1969)

[14] PH.NOVERRAZ, *Fonctions plurisousharmoniques et analytiques sur les espaces vect. top.*, Ann. Inst. Fourier 19 (1969)

[15] M.SCHOTTENLOHER, *Analytische fortzetzung in Banachräumen*, Math. Ann. (à paraître)

SOME RECENT RESULTS ON TOPOLOGICAL VECTOR SPACES

by

M. A. Dostal

This article reproduces loosely the course entitled "Homological algebra and topological vector spaces [1]" given at the Symposium of Analysis in Recife in July 1972. As the change in the title indicates, this is not an exact account of the seven-hour course whose purpose was to convey, if not the main results, then at least the flavor of a new field in TVS, which originated not long ago in the seminar of D.A.Raĭkov in Moscow. Actually, the present article gives only an introduction to this new approach to TVS (chapters 3 and 4). However, some other topics, which were merely touched upon in the course, are treated here in more details (chapters 1 and 2). The main reason for not concentrating entirely on the homological methods is the limitation of space: indeed, a more thorough treatement of these topics would necessitate a detailed discussion of several homological concepts which differ from their counterparts in the standard homological algebra (cf. [67]), and this would make the article too long. However, any reader wishing to penetrate deeper into the applications of homological algebra should consult the excellent survey article [67] written by one of the main contributors to this field, V.P. Palamodov. (An English translation of [67] has just become available.) Finally, it should be mentioned that the choice of the material presented below was also influenced by the desire to acquaint the reader with certain results which were originally published in East European journals and thus not always easily accessible.

Let us give a short survey of the contents.

Chapter 1: After recalling the concept of full completeness and related results, some of which are of recent date, the chapter continues with a brief discussion of De Wildes's spaces with webs. The main objective is to prove a homomorphism theorem due to De Wilde (cf. [14] and (1,23) below). This theorem is very useful in applications, and moreover, a short and self-contained proof of it has apparently not been published.

---

1) This term will be abbreviated in the sequel as TVS.

Chapter 2 discusses some of the recent results on the open mapping theorem in inductive limits of Fréchet spaces. Emphasis was put on the work of V.Pták. Needless to say, a complete account of these questions would have to cover important contributions of D. A. Raĭkov, W. Słowikowski and others. However, this would again go far beyond the scope of the present article.

Chapter 3 contains an introduction to the so-called free locally convex spaces and related topics. Although relatively old (A. A. Markov, 1941) and of natural character, the concept of a free locally convex space has never received much attention in the theory of TVS. Nonetheless, it periodically reappeared in papers of different authors (Cf. [44, 45, 57, 78]), thus demonstrating its rightful place in functional analysis. Moreover, this concept and its modifications represent one of the non-trivial and less explored links between general topology and functional analysis.

Chapter 4 deals with various properties of the category of locally convex spaces. In particular, it is shown there that this category possesses sufficiently many injectives, but not sufficiently many projectives [2]. This is, of course, of crucial importance for applications of homological algebra to this category. The chapter ends with the proof of an interesting result of V. P. Palamodov (cf. (4,22) below).

Originally, it was planned to add a chapter containing homological proofs of some of the statements of Chapter 2. Only this would demonstrate how powerful a tool we gain in the use of homological methods. Moreover, it would justify our choice of the material, which otherwise might seem to have been selected mainly on the basis of personal preferences of the author. A detailed treatement of these topics will be found in a larger publication which we are preparing.

In conclusion the author wishes to thank various participants in the conferences at Recife and Rio de Janeiro for their helpful comments; these will be recalled at the corresponding places in the text. However the author's special thanks go to M. De Wilde for his most valuable help in editing sections (1,12)-(1,24), and furthermore to M.J.Kascic, Jr., for carefully checking the manuscript.

---

2) This result is very recent and is due to V.A.Geĭler [30]. Moreover it shows the usefulness of free locally convex spaces.

§ 0 - NOTATION

Ab: the category of Abelian groups

(B): the class of all Banach spaces

cv. A: the (absolutely) convex hull of the set A

Ens: the category of sets

E_b': the strong dual of a TVS E

(E,t): the topological space with the topology t

(F): the class of all Fréchet spaces

$\mathbb{K}$: either one of the two fields: $\mathbb{R}$ (= the real number field); $\mathbb{C}$ (= the complex number field)

$\mathbb{K}^A$ (resp. $\mathbb{K}^{(A)}$), where A is a cardinal: the TVS (over $\mathbb{K}$) defined as the cartesian product (direct sum resp.) of A copies of $\mathbb{K}$. If A denotes a set, then we define $\mathbb{K}^{(A)} = \mathbb{K}^{(\mathrm{card}(A))}$ and $\mathbb{K}^A = \mathbb{K}^{\mathrm{card}(A)}$.

$\ell.c.$: locally convex

(LB) (resp. (LF)): the class of strict inductive limits of sequences of Banach spaces (resp. Fréchet spaces) (cf. [5])

$\mathbb{N}$: the set of positive integers

P(R): Let E be a $\ell.c.$ space and R a subspace of E. Then for any $u \in E'$, P(R)u denotes the restriction of u to R (cf. [76,77])

Spec E: the family of all continuous seminorms on a given $\ell.c.$ space E

sup T: the coarsest topology finer than each element of the family T

$\mathcal{T}_1 \subseteq \mathcal{T}_2$: the topology $\mathcal{T}_2$ is finer than the topology $\mathcal{T}_1$

TCR: the category of completely regular topological spaces [1)]

---

1) including all non-Hausdorff spaces as well.

TLC: the category of ℓ.c. spaces [1] over $\mathbb{K}$ [2]. Morphisms in this category are all continuous linear mappings.

$\mathcal{U}(E)$: the family of *all* neighborhoods of the origin in a ℓ.c. space E

$\hat{\mathcal{U}}(E)$: the family of all closed absolutely convex sets $U \in \mathcal{U}(E)$

UN: the category of uniform spaces

V: the category of vector spaces over $\mathbb{K}$ [2]

Let C be a category and X and Y objects in C. Then 1_X denotes the identity morphism of X in the category C. Similarly, $X \cong Y$ means that objects X and Y are isomorphic in C.

Let $T \in L(E,F)$ where E and F are ℓ.c. spaces. Then TE denotes the range of T equipped with the topology induced from F. T is said to be a *homomorphisms* if the mapping $T: E \longrightarrow TE$ is open. Similarly one defines *weak homomorphisms* (cf. [5]). Furthermore, we say T is *injective* (*surjective* resp.) instead of saying that T is one-to-one (onto resp.).

Let (F,G) be a pair of ℓ.c. spaces in duality. The canonical bilinear form defining this duality is denoted by $\langle\cdot,\cdot\rangle_{F,G}$. If G = F' we also write $\langle\cdot,\cdot\rangle_F$ or just simply $\langle\cdot,\cdot\rangle$.

$A^o \subseteq R'$: Let R be a subspace of a ℓ.c. space F and $A \subseteq R$. Then $A^o \subseteq R'$ means that the polar of A is taken in R' and not in F'. $F_j \longrightarrow F$ means the same as $F = \lim_j \operatorname{ind} F_j$, i.e. a ℓ.c. space F is a strict inductive limit of Fréchet spaces F_j. $\{F_j\}$ is then said to be a *defining sequence* for F.

A ℓ.c. space E is called *σ-infrabarrelled* (or also σ-quasi-barrelled or σ-evaluable) if all countable bounded subsets of E'_b are equicontinuous (cf. [68]). For all other definitions and facts on TVS the reader should consult standard monographs such as [5, 22, 29, 36, 37, 41, 49, 53, 68, 86, 89, 106, 107]. The necessary material on categorical algebra can be found in [8, 28].

---

2) It is assumed that $\mathbb{K}$ is one and the same for all spaces in this category.

§ 1 - A HOMOMORPHISM THEOREM

Throughout this chapter all $\ell.c.$ spaces are assumed to be Hausdorff.

The central question of functional analysis in TVS is to decide whether a given $T \in L(E,F)$ is a homomorphism or not. Another problem closely related to the preceding one is to find under what conditions a linear mapping $T: F \longrightarrow E$ with closed graph is continuous. These two problems are of primary importance for both the theory of TVS and its applications in other parts of analysis.

If E and F are Fréchet spaces, then it is known that $T \in L(E,F)$ is a homomorphism whenever $TE = F$; and, under the same hypothesis on E and F, the answer to the second question is always in the affirmative [1]. These two classical results, known as Banach's open-mapping and closed-graph theorems, turned out to be so useful in applications that several authors started to analyze their structure and succeeded in extending their validity far beyond the class of Fréchet spaces. The earlier stages of this research were dominated by the work of V.Pták, A. P.Robertson and W.Robertson. [1]

When we try to generalize Banach's theorems to non-metrizable spaces, we quickly realize that their straightforward generalization to complete spaces is not possible. Indeed, there exist complete non metrizable but barrelled spaces in which these theorems do not hold [1].

---

1) cf. [22, 37, 71, 86] and their respective bibliographies.

Hence in order to find a proper generalization of Banach's theorems, one must strengthen the hypothesis of completeness. However to this purpose a more explicit characterization of completeness is necessary. Such a characterization (in terms of duality) was found by Grothendieck in 1950:

(1,1) THEOREM [31]: *The completion* $\hat{E}$ *of a l.c. space* E *is* (algebraically) *isomorphic to the vector space of all linear functionals on* E' *which are* $\sigma(E', E)$-*continuous on every equicontinuous subset of* E'.

(1,2) COROLLARY: *If* $(E, \mathcal{T}_1)$, $(E, \mathcal{T}_2)$ *are l.c. spaces such that* $\mathcal{T}_1 \subseteq \mathcal{T}_2$ *and* $(E, \mathcal{T}_1)' = (E, \mathcal{T}_2)'$, *then* $(E, \mathcal{T}_2)^{\wedge}$ *is (algebraically) isomorphic to a subspace of* $(E, \mathcal{T}_1)^{\wedge}$.

(1,3) DEFINITION: Let E be a *l.c.* space and Q a subset of E'. Then Q is said to be *almost closed* if for each $U \in \mathcal{U}(E)$, the set $Q \cap U^{o}$ is closed in the relative $\sigma(E', E)$-topology on U^{o}.

(1,4) COROLLARY: *A l.c. space* E *is complete iff for every hyperplane* Q *in* E' *one has:*

(a.c.) Q *is* $\sigma(E', E)$-*closed iff* Q *is almost closed.*

This corollary shows that a natural way of strengthening the concept of a complete space is described in the following:

(1,5) DEFINITION: A *l.c.* space E is called

(B_r) B_r-*complete*, if *(a.c.)* holds for all subspaces $Q \subseteq E'$ which are dense in E' for the topology $\sigma(E', E)$;

(*f.c.*) B-*complete* or *fully complete*, if *(a.c.)* holds for all subspaces Q of E';

(*hyp*) *hypercomplete*, if (*a.c.*) holds for all absolutely convex subsets Q of E';

(K-Š) a *Kreĭn-Šmulian space* , if (*a.c.*) holds for all convex subsets Q of E'.

(*her*) *hereditary complete*, if E/N is complete for an arbitrary closed subspace N of E.

If (c) denotes the class of all complete spaces, then the relationships between the above types of completeness can be summarized as follows:

(1,6) "THEOREM":

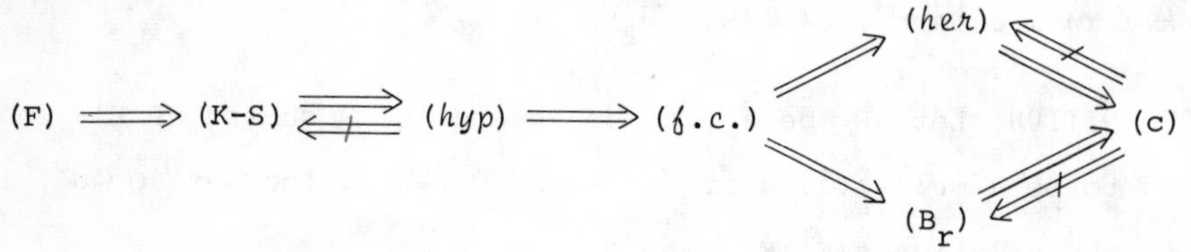

"PROOF" [2]: That (F) $\Longrightarrow$ (K-Š) was first established by M. Kreĭn and V.Šmulian for the case of Banach spaces in [54] (cf. also [36]). Fully complete and B_r-complete spaces were introduced by V.Pták [69, 61] who studied them in great detail. Among other facts Pták showed that (c) $\not\Longrightarrow$ (B_r). H.S.Collins [9, 10] studied independently the class of fully complete spaces and which that a closed subspace and a quotient space of a fully complete space are necessarily fully complete; in particular, (*f*.c.) $\Longrightarrow$ (*her*). This is interesting, for it was known for a long time that (c) $\not\Longrightarrow$ (*her*) (cf. G.Köthe [50]). On

---

2) The missing implications in the above diagram indicate questions to our best knowledge are still open.

the other hand, if we had $(B_r) \Longrightarrow (her)$, then this would easily imply $(f.c.) \Longleftrightarrow (B_r)$ (Cf. [22], p.541, Prop. 8.10.10). This, however, still seems to be an open problem (cf. the remark following this paragraph). Hypercomplete spaces were introduced by J.L. Kelley [47]; an example showing that $(hyp) \not\Longrightarrow$ (K-Š) can be found in [49], 18H, p.178. Later it turned out that the space $\mathcal{D}'$ of Schwartz distributions on the real line is not fully complete. This was first observed by D.A.Raikov [81] on the basis of a counterexample due to W. Slowikowski [93] (cf. also [34, 67]). This result can be further adapted to show that $\mathcal{D}'$ is not hereditary complete ([21, 34, 67]).

REMARK: According to an oral communication of Prof. W.H.Summers, D. van Dulst has recently proved that the space $\mathcal{D}'(\Omega)$ (Ω arbitrary open subset of $\mathbb{R}^n$) is B_r-complete. This result thus would end a long search for a space which would be B_r-complete but not fully complete. In view of what was said above this would also imply that $(B_r) \not\Longrightarrow (her)$ [3]. Another unsolved problem has been the full completeness of the space $\mathcal{D}$. However it is mentioned in the footnote on p.60 of [67] that O.G.Smolyanov has recently disproved the hereditary completeness of $\mathcal{D}$ (no bibl. reference is given). Hence $\mathcal{D} \notin (f.c.)$. From all these results (and open problems) one can draw two conclusions: First, the Schwartz spaces $\mathcal{D}(\Omega)$ and $\mathcal{D}'(\Omega)$ considered as $\ell.c.$ spaces exhibit rather pathological properties. Second, in concrete cases it can be very difficult to verify whether the space under consideration is fully complete, B_r-complete, etc. (For more details and further references, cf. [11, 20, 21, 22, 36, 37, 71, 86, 89].)

---

[3] It is stated in [37] on p.56 that B_r-complete spaces are hereditary B_r-complete. According to the above this would mean $(f.c.) \Longleftrightarrow (B_r)$ and therefore would contradict the result of van Dulst.

To relate the above concepts of completeness to the open-mapping and closed-graph theorems we first need a simple definition:

(1,7) DEFINITION: Let E and F be ℓ.c. spaces and T: E $\longrightarrow$ F a linear mapping. Then T is called *almost open* (*almost continuous* resp.), if for each $U \in \mathcal{U}(E)$ ($V \in \mathcal{U}(F)$ resp.), we have $\overline{TU} \in \mathcal{U}(F)$ ($\overline{T^{-1}V} \in \mathcal{U}(E)$ resp.).

The following result is due to Pták ([71]; cf. also [36]):

(1,8) THEOREM: *A ℓ.c. space E is fully complete iff any almost open continuous linear mapping of E into any ℓ.c. space F is open.*

(1,9) REMARK: If E is an arbitrary ℓ.c. space and F barrelled, then obviously every linear mapping T of E onto F is almost open. And similarly, every linear mapping S: F $\longrightarrow$ E is almost continuous. Hence (1,8) yields immediately the following generalization of the Banach open-mapping theorem:

Let E be fully complete and F barrelled. Then every linear continuous mapping of E onto F is open.

This can be further strengthened as follows:

(1,10) THEOREM [71]: *Let E be B_r-complete and F barrelled. Then every linear mapping of E onto F with closed graph is open.*

Similarly, one can establish the corresponding version of the closed-graph theorem:

(1,11) THEOREM [71, 85, 89]: *Let E be B_r-complete and F barrelled. Then every linear mapping of F into E with the closed graph is continuous.*

For further results of this kind the reader is referred to [37, 71, 86, 87].

These results, however elegant and general, have basically two inconvenient features: First, the class of fully complete spaces does not possess good permanence properties (cf. [19, 32, 98]). Thus it was shown by W.H.Summers [98][4)] that a product of two Kreĭn-Šmulian spaces need not be fully complete. Second, relatively few fully complete spaces are known. This is due to the fact that for a concrete space E, it is usually difficult to decide whether E satisfies one of the conditions in (1,5) (cf. the case of the Schwartz spaces $\mathcal{D}(\Omega)$ and $\mathcal{D}'(\Omega)$ cf. also [11, 98, 110]).

One way of overcoming these difficulties can be formulated as follows:

PROBLEM: Consider the open-mapping theorem (1,10). Find a version of this theorem, in which the full completeness of E would be replaced by a less stringent condition, while more restrictions would be put on F. And similarly for the close-graph theorem.

In a more concrete form, this problem was formulated first by A. Grothendieck in [33]. Results of this kind were later found by several authors including A.Grothendieck (1955; [33]), W.Słowikowski (1961; [91, 92]), D.A.Raĭkov (1966; [79,80]), L.Schwartz (1966; [90]), A. Martineau (1966; [60, 61]; cf. also [106]), M. De Wilde (in a series of papers

---

4) I wish to thank Prof. J.B.Prolla for bringing this article to my attention. For a detailed discussion of the result announced in |98| and other related facts, cf. [110].

starting from 1967; cf. [15]), W.Robertson (1972; [88]) and many others. In what follows we shall discuss only one result of this kind, due to De Wilde (cf. (1,17)). This in turn will imply a homomorphism theorem (1,23) which will be useful in the next chapter.

(1,12) DEFINITION [5]: Let E be a ℓ.c. space. Then a *web* in E is an arbitrary family $R = \{e_{n_1 \ldots n_k} \subset E : k, n_1, \ldots, n_k = 1,2,\ldots\}$ such that

(a) $E = \bigcup_{n_1=1}^{\infty} e_{n_1}$; $e_{n_1} = \bigcup_{n_2=1}^{\infty} e_{n_1 n_2}$; ...; $e_{n_1 \ldots n_k} = \bigcup_{n_{k+1}=1}^{\infty} e_{n_1 \ldots n_{k+1}}$; ...

for all positive integers $k, n_1, n_2, \ldots, n_k$.

R is said to be a *strict web*, if all sets in (a) are absolutely convex and

(b) for every sequence $\{n_k\} \subset \mathbb{N}$ there exists a sequence of $\lambda_k \neq 0$ $(k \geq 1)$ such that whenever $x_k \in e_{n_1 \ldots n_k}$ for $k = 1,2,\ldots$, the series $\sum_{k=1}^{\infty} \lambda_k x_k$ converges in E and $\sum_{k=k_o}^{\infty} \lambda_k x_k \in e_{n_1 \ldots n_{k_o}}$ for all k_o.

(1,13) REMARK: Assume that a sequentially complete space E has a web R all of whose members are absolutely convex closed sets. Then R will be a strict web provided the following holds:

(b') for arbitrary sequence $\{n_k\}_{k \geq 1} \subset \mathbb{N}$ there are $\lambda_k' \neq 0$ $(k \geq 1)$ such that for every sequence $x_k \in e_{n_1 \ldots n_k}$ $(k \geq 1)$, the sequence $\{\lambda_k' x_k\}_{k \geq 1}$ is bounded in E. (Indeed, then it suffices to set $\lambda_k = 2^{-k} \lambda_k'$ and condition (b) will be satisfied.)

---

5) All results in the rest of this chapter (except (1,20) and (1,21)) are due to De Wilde, cf. [15].

(1,14) PROPOSITION: *If* E *admits a strict web* R, *then every sequentially closed subspace* E_1 *of* E *has a strict web, namely*

$$R_1 = \{E_1 \cap e_{n_1 \ldots n_k} : k, n_1, \ldots, n_k \in \mathbb{N}\}.$$

PROOF is obvious.

Webs become particularly transparent in Fréchet spaces:

(1,15) PROPOSITION: *Every l.c. metrizable space* E *has a web. In particular, every Fréchet space* E *has a strict web.*

PROOF: Let $\{p_n\}_{n \geq 1}$ be a basis of *Spec* E. We may assume $p_1 \leq p_2 \leq \cdots$. Set $b_{p_k}(\varepsilon) = \{x \in E : p_k(x) \leq \varepsilon\}$. The sets

$$e_{n_1 \ldots n_k} = b_{p_1}(n_1) \cap \ldots \cap b_{p_k}(n_k); \quad k, n_1, \ldots, n_k \in \mathbb{N}$$

obviously define a web. Then it suffices to apply (1,13).

(1,16) PROPOSITION: *Let* E *be an inductive limit of l.c. metrizable spaces* E_i $(i = 1,2,\ldots)$. *Then the strong dual* E'_b *has a strict web.*

PROOF: For each $i = 1,2,\ldots$, let $\{U_{i,k}\}_{k \geq 1}$ be a basis of $\hat{\mathcal{U}}(E_i)$ such that $2U_{i,k+1} \subset U_{i,k}$ for all $k \geq 1$. Set

(1) $$e_{n_1} = U^{\circ}_{1,n_1}; \quad e_{n_1 n_2} = U^{\circ}_{1,n_1} \cap U^{\circ}_{2,n_2}; \quad \ldots;$$

$$e_{n_1 \ldots n_k} = U^{\circ}_{1,n_1} \cap \ldots \cap U^{\circ}_{k,n_k}; \text{ for all } k, n_1, \ldots, n_k \in \mathbb{N}.$$

Then the family R of all sets in (1) is clearly a web in E'. Choose arbitrarily the sequences $\{n_k\}_{k \geq 1}$ and $x_k \in e_{n_1 \ldots n_k}$ $(k \geq 1)$. Since

$$V = \{x_k\}_{k \geq 1} \subset U^{\circ}_{1,n_1},$$

V is equicontinuous in E_1'. Similarly, $\{x_k\}_{k\geq 2} \subset U^o_{2,n_2}$ shows that V is equicontinuous en E_2'. Hence V is equicontinuous in every E_k' and thus also in E'. If we choose $\lambda_k' = 2^{-k}$, the proposition then follows from (1,13).

Now we are able to prove a version of the open-mapping theorem:

(1,17) THEOREM: *Let* E *be a Fréchet space and* F *a space with a strict web. Let* T *be a linear mapping defined on a subspace* F_1 *of* F *and with values in* E. *Furthermore, let the graph of* T *be sequentially closed in* $F\times E$ *and* $TF_1 = E$. *Then there exists a sequence* $\{n_k\}_{k\geq 1}$ *such that* $Te_{n_1\ldots n_k} \in \mathcal{U}(E)$ *for all* $k \geq 1$.

PROOF: It follows from our assumptions that

$$(2) \qquad E = TF_1 = \bigcup_{n_1} Te_{n_1}; \quad Te_{n_1} = \bigcup_{n_2} Te_{n_1 n_2}; \ \ldots, \text{ etc.}$$

Since E is a Fréchet space, there exists n_1 such that Te_{n_1} is of second category in E. Similarly, for some n_2, $Te_{n_1 n_2}$ must be of second category in E, etc. Thus we obtain a sequence $\{n_k\}_{k\geq 1}$ such that each $Te_{n_1\ldots n_k}$ is a set of second category in E, hence

$$\overline{Te_{n_1\ldots n_k}} \in \hat{\mathcal{U}}(E).$$

Take a sequence $\{\lambda_k\}_{k\geq 1}$ associated with the sequence $\{n_k\}$ according to (1,12)(b). Furthermore, choose $U_k \in \mathcal{U}(E)$ so that $\{U_k\}_{k\geq 1}$ is a basis of $\mathcal{U}(E)$ and

$$(3) \qquad U_k \subset \lambda_k \overline{Te_{n_1\ldots n_k}}.$$

Fix arbitrarily k_o and $x_{k_o} \in U_{k_o}$. By (3),

$$U_{k_o} \subset \lambda_{k_o} Te_{n_1 \ldots n_{k_o}} + U_{k_o+1} .$$

Hence for some $y_{k_o} \in e_{n_1 \ldots n_{k_o}}$ and $x_{k_o+1} \in U_{k_o+1}$,

(4) $$x_{k_o} = \lambda_{k_o} Ty_{k_o} + x_{k_o+1} .$$

Similarly we obtain $x_{k_o+2} \in U_{k_o+2}$ and $y_{k_o+1} \in e_{n_1 \ldots n_{k_o+1}}$ such that

(5) $$x_{k_o+1} = \lambda_{k_o+1} Ty_{k_o+1} + x_{k_o+2} ,$$

etc. Finally, we obtain sequences $x_k \in U_k$, $y_k \in e_{n_1 \ldots n_k}$ $(k \geq k_o)$ for which $x_k = \lambda_k Ty_k + x_{k+1}$. In particular, by (4),

(6) $$x_{k_o} = \sum_{k=k_o}^{N} \lambda_k Ty_k + x_{N+1} \quad (N = k_o+1, \ldots).$$

By our choice of U_k, $x_k \longrightarrow 0$, thus by (6),

(7) $$x_{k_o} = \sum_{k=k_o}^{\infty} \lambda_k Ty_k .$$

On the other hand, by (1,12)(b),

(8) $$y = \sum_{k=k_o}^{\infty} \lambda_k y_k \in e_{n_1 \ldots n_{k_o}} .$$

The graph of T being sequentially closed in $F \times E$, we get from (7) and (8), that $y \in F_1$ and $Ty = x_{k_o}$. Thus we have proved that

$$U_{k_o} \subset Te_{n_1 \ldots n_{k_o}} .$$

(1,18) COROLLARY: *Let* E, F *and* T *be as in the previous theorem. If* $x_n \longrightarrow 0$ *in* E, *then there are* $y_n \in F$ $(n \geq 1)$ *such that* $y_n \longrightarrow 0$ *and* $Ty_n = x_n$ *for all* n.

PROOF: By (1,17) there are $U_k \in \mathcal{U}(E)$ such that

(9)
$$U_k \subset \lambda_k Te_{n_1 \ldots n_k} \qquad (k \geq 1).$$

If $x_m \longrightarrow 0$ in E, then there exists a sequence $m_k \nearrow \infty$ such that $x_m \in U_k$ for all $m \geq m_k$ and $k \geq 1$. If $m_k \leq m < m_{k+1}$ then by (9), $x_m = Ty_m$ for some $y_m \in \lambda_m e_{n_1 \ldots n_k}$. Assume that $y_m \not\longrightarrow 0$. Hence there exists a $U \in \mathcal{U}(F)$ such that $y_{m_i'} \notin U$ for infinitely many m_i'. Moreover, we can assume that each interval $[m_k, m_{k+1})$ contains at most one m_i'. By (1,12)(b), the series $\sum y_{m_i'}$ converges. Whence $y_{m_i'} \longrightarrow 0$, a contradiction.

(1,19) In applications the following situation frequently occurs: Let $T \in L(E,F)$ be a mapping which is not necessarily onto, but is open in some weakened sense (e.g., T is a weak homomorphism). Under what conditions on E, F and T, can one conclude that T is a homomorphism? Criteria of this kind are usually called homomorphism theorems. Although they are related to open-mapping theorems, they cannot always be reduced to this type of theorems (cf. §2). In the rest of this chapter we shall discuss the case when T is a weak homomorphism. The next theorem summarizes some well-known facts about weak homomorphisms. For proofs and further discussion, cf. [5, 7, 73, 107].

(1,20) THEOREM: *Let* $T \in L(E,F)$. *Then the following implications hold:*

(a) T *is a homomorphism* $\Longrightarrow$ (b) $\Longleftrightarrow \ldots \Longleftrightarrow$ (g),

where

(b) T *is a weak homomorphism;*

(c) $T'F'$ *is* $\sigma(E',E)$*-closed in* E';

(d) $T'F' = (T^{-1}(0))^{\circ}$;

(e) $T'F' \supseteq (T^{-1}(0))^{\circ}$;

(f) *if* P *is a closed subspace of* E *such that* $P \supseteq T^{-1}(0)$, TP *is closed in* TE.

(g) *the same condition for closed hyperplanes* P.

The following is a classical homomorphism theorem:

(1,21) THEOREM: *Let* E *and* F *be Fréchet spaces. Then all conditions* (a), (b),..., (g) *in* (1,20) *are equivalent to the condition:*

(h) TE *is closed in* F.

For applications the assumption that E and F are Fréchet spaces is often too restrictive. To prove a different homomorphism theorem, we first need a simple lemma:

(1,22) LEMMA: *Let* $T \in L(E,F)$ *be a weak homomorphism with the following property:*

(10) *For every* $U \in \hat{\mathcal{U}}(E)$, *there exists* $V \in \hat{\mathcal{U}}(F)$ *such that* $U^{\circ} \cap T'F' \subseteq T'V^{\circ}$.

Then T *is a homomorphism.*

PROOF: Let U and V be as in (10). Then (1,20) combined with simple properties of polar sets implies:

$$(11)\quad \begin{cases} T^{-1}V = (T'V^{\circ})^{\circ} \subseteq (U^{\circ} \cap T'F')^{\circ} = \overline{cv}(U^{\circ\circ} \cup (T'F')^{\circ}) \\ \subseteq \overline{U + (T'F')^{\circ}} \subseteq 2U + (T'F')^{\circ} \\ = 2U + T^{-1}(0). \end{cases}$$

Hence for $V_1 = (1/2)V$, relation (11) yields $TU \supseteq V_1 \cap TE$, which proves the lemma.

(1,23) THEOREM: *Let* E *be a Schwartz space and* F *a* σ-*infrabarrelled l.c. space whose strong dual* $F' = F'_b$ *admits a strict web. Then* $T \in L(E,F)$ *is a homomorphism if (and only if)* T *is a weak homomorphism.*

PROOF: Let $U \in \hat{\mathcal{U}}(E)$ be arbitrary. According to the dual description of a Schwartz space (cf. [36], Prop. 5, p.277), there is a $W \in \hat{\mathcal{U}}(E)$ such that U° is compact in the Banach space $H = (E')_{W^{\circ}}$. Denote by H_1 (H_2 resp.) the space $T'F'$ equipped with the topology induced from H (E'_b resp.). By (1,20), $U^{\circ} \cap T'F'$ is compact in the Banach space H_1. By [29] b), p.82, H_1 contains a sequence $X = \{x_m\}_{m\geq 1}$, $x_m \longrightarrow 0$, such that $U^{\circ} \cap T'F' \subset \overline{cv}\, X$. Obviously, $T' \in L(F', H_2)$, in particular, the graph of T' is closed in $F' \times H_1$. By virtue of (1,18), there exists $Y = \{y_m\}_{m\geq 1} \subset F'$, $y_m \longrightarrow 0$, $T'y_m = x_m$ ($\forall\, m$). By our assumption on F, $Y \subset V^{\circ}$ for some $V \in \hat{\mathcal{U}}(F)$. Therefore, $X = T'Y \subset T'V^{\circ}$. Since the set $T'V^{\circ}$ is clearly absolutely convex and $\sigma(E',E)$-closed, we obtain $U^{\circ} \cap T'F' \subset \overline{cv}\, X \subset T'V^{\circ}$. Condition (10) of (1,22) is verified and the theorem follows.

(1,24) COROLLARY: *Let* E *be a Schwartz space and* F *a countable inductive limit of l.c. metrizable spaces. Then every weak homomorphism* $T \in L(E,F)$ *is also a homomorphism in the strong topologies.*

The last statement turns out to be very useful in applications (cf. [3, 35]). For further development of De Wilde's theory, the reader should consult [15, 88, 99].

§ 2 - OPENNESS IN (LF)-SPACES

Throughout this chapter all $\ell.c.$ spaces are assumed to be Hausdorff and all inductive limits are strict (cf. §0).

In what follows we shall be concerned with a problem which originated in the theory of convolution operators. It can be formulated roughly as follows:

DIVISION PROBLEM: Let X and Y be spaces of (generalized) distributions. Consider a convolution operator $S: X \longrightarrow Y$. When is the equation $S*u = v$ solvable in $u \in S$ for arbitrary $v \in Y$?

In applications, the spaces X and Y are usually subspaces of the Schwartz space $\mathfrak{D}' = \mathfrak{D}'(\mathbb{R}^n)$, such as $\mathfrak{E}'(\Omega)$, $\mathfrak{D}'(\Omega)$, or $\mathfrak{D}'_F(\Omega)$, where Ω is an open subset of $\mathbb{R}^n$; or X and Y can be spaces of analytic functionals or other spaces of more general "distributions". The operator S can be either a linear partial differential operator with constant coefficients or a difference-differential operator or a convolution operator of a more complicated type. For a thorough discussion of various concrete division problems the reader is referred to [2, 3, 17, 56, 23, 24, 25, 35, 105].

In this chapter we shall discuss the functional analytic aspect of the division problem. In order to find a proper abstract formulation of the division problem, let us recall that in concrete cases S is usually the adjoint of an injective convolution operator $T: E \longrightarrow F$ where E and F are (LF)-spaces such that $E' = Y$ and $F' = X$. Then our problem can be formulated as follows:

(2,1) PROBLEM: *Let* E *and* F *be* (LF)*-spaces and* T *an injective linear continuous mapping of* E *into* F. *Under which additional conditions on* E, F *and* T *is the adjoint operator* T' *surjective, i. e.* (cf. (1,20)), *when is* T *a weak homomorphism?*

(2,2) REMARK: To solve problem (2,1), it obviously suffices to find conditions which would imply that T is a (strong) homomorphism. If E is a Schwartz space, then by (1,24) the last condition is also necessary. (Since in many concrete cases the spaces under consideration are nuclear, this remark can be very useful, cf. [3]).

It should be pointed out that since in the formulation of (2,1) we did not use any specific properties of convolution operators, problem (2,1) is still too general. One possible restriction of T is described in the next definition:

(2,3) DEFINITION: Let $E, F \in (LF)$, $F = \lim_j \operatorname{ind} F_j$ and $T \in L(E,F)$. Then T is called *sequentially open*, if the range TE of T is sequentially closed in F, i.e.

(a) $TE \cap F_j$ is closed for all $j = 1,2,\ldots$.

(2,4) DEFINITION: Let R and W be two absolutely convex subsets of $F = \lim_j \operatorname{ind} F_j \in (LF)$, $W \subseteq R$. Then W is called a *sequential neighborhood* in R, if there are $V_j \in \mathcal{U}(F_j)$ such that for every $j = 1,2,\ldots$,

(I,j) $$V_j \cap R \cap F_j \subseteq W.$$

The next lemma due to Pták [73] elucidates the meaning of definition (2,3):

(2,5) LEMMA: *Let* E, F *and* T *be as in* (2,3). *Then condition* (a) *of* (2,3) *is equivalent to either one of the following two conditions:*

(b) *There exists a defining sequence* $\tilde{E}_j \longrightarrow E$ *such that for all* $j = 1,2,\ldots,$ $T\tilde{E}_j$ *is closed and* $T\tilde{E}_j = TE \cap F_j$.

(c) *There exists a defining sequence* $\tilde{E}_j \longrightarrow E$ *with property* (b) *such that* T *restricted to each* $\tilde{E}_j$ *is a homomorphism, and for every* $U \in \mathcal{U}(E)$, TU *is a sequential neighborhood in* TE.

(Since this lemma is not essential for our discussion, the proof will be omitted; cf. [73].)

(2,6) REMARK: (i) Condition (c) justifies the terminology of (2,3).

(ii) It follows from (2,4) that if $R = F$, then W is a sequential neighborhood iff $W \in \mathcal{U}(F)$.

(iii) If T is surjective, then by (2,3) T is sequentially open. Hence by (c) and (ii), T is open. We have thus obtained the open mapping theorem for (LF)-spaces, which is due to J.Dieudonné and L. Schwartz [16]: *Let* $E, F \in$ (LF) *and* $T \in L(E,F)$ *be surjective. Then* T *is a homomorphism.*

By the last remark, the difference between openness and sequential openness arises only when T is not surjective. Comparing this fact with (2,1) and (2,2) we see that we are again dealing with a problem similar to the one discussed in (1,9): *Let* $E, F \in$ (LF) *and let* $T \in L(E,F)$ *be sequentially open and injective, but not necessarily surjective. When is* T *a homomorphism (resp. a weak homomorphism)?*

First, let us show how this problem can be reduced to the study of subspaces of F – or more exactly – to the study of the "location in F" of sequentially closed subspaces of F.

(2,7) Consider an (LF)-space $F = (F, \nu) = \lim_j \mathrm{ind}\, F_j$ and a sequentially closed subspace $R \subseteq F$. Set $R_j = R \cap F_j$. Then $R_j \in (F)$, hence R can be equipped with the topology ι_R of the inductive limit $\lim_j \mathrm{ind}\, R_j$. Let ν_R be the subspace topology on R, i.e. the topology induced on R by ν. Set $R_\nu = (R, \nu_R)$ and $R_\iota = (R, \iota_R)$. Let R'_ν and R'_ι be the corresponding (strong) duals. Since obviously $\iota_R \supseteq \nu_R$, the identity mapping

(1) $$I: R_\iota \longrightarrow R_\nu$$

is continuous. Therefore the adjoint mapping

(2) $$I': R'_\nu \longrightarrow R'_\iota$$

is continuous and injective, and its image $I'R'_\nu$ is weakly dense in R'_ι.

(2,8) DEFINITION: R is called an (LF)-*subspace* (of F), if the mapping I is a homomorphism (hence an isomorphism), i.e. if $R_\iota = R_\nu$. Similarly, R is said to be *well-located* (in F), if I is a weak homomorphism, i.e. if I' is surjective: $R'_\iota = R'_\nu$.

(2,9) REMARK: Every (LF)-subspace R of F is clearly a closed well-located Mackey-Arens subspace [1] of F. Moreover, $\mathcal{U}(R_\iota)$ is the family of all sequential neighborhoods on R. Hence in order to show that a given R is an (LF)-subspace, we must have some criterion for a sequential neighborhood to be a neighborhood (in the topology ν_R). The

---

1) i.e. $\nu_R = \tau(R, R'_\nu)$.

following lemma represents such a criterion. It is a simple but efficient device which goes back to [56]. Later it was used by Hörmander [35] in his study of strong P-convexity. In its present simplified form it is due to Pták [73]:

(2,10) LEMMA: *Let* F, R *and* W *be as in* (2,4). *Suppose that for any* $j = 1,2,\ldots$, *any* $\varepsilon \in (0,1)$ *and arbitrary* $V_j \in \mathcal{U}(F)$ *satisfying condition* (I,j) *of* (2,4), *there exists* $V_{j+1} \in \mathcal{U}(F)$ *satisfying condition* (I, j+1) *and*

(II,j) $$(1 - \varepsilon)(V_j \cap F_j) \subseteq V_{j+1} \subseteq V_j.$$

Then $W \in \mathcal{U}(R_\nu)$.

PROOF: Fix $\varepsilon_j \in (0,1)$ $(j = 1,2,\ldots)$ arbitrarily but so that $\eta = \prod_1^\infty (1 - \varepsilon_j)$ converges and is positive. Let $V_1 \in \mathcal{U}(F)$ be such that (I,1) holds. Therefore, there is a $V_2 \in \mathcal{U}(F)$ satisfying (I, 2) and (II, 1) (with $\varepsilon = \varepsilon_1$). Next we choose $V_3 \in \mathcal{U}(F)$ satisfying (I, 3) and (II, 2) (with $\varepsilon = \varepsilon_2$), etc. Set $V = \bigcap_j V_j$ and $\nu_j = \prod_{k \geq j} (1-\varepsilon_k)$. Then by (II,j),

$$\eta(V_j \cap F_j) \subseteq \eta_j(V_j \cap F_j) \subseteq \ldots \subseteq \eta_{j+p}(V_{j+p} \cap F_{j+p}) \subseteq V_{j+p},$$

for $p = 1,2,\ldots$. This together with the second inclusion in (II, j) gives $\eta(V_j \cap F_j) \subseteq V$ $(j = 1,2,\ldots)$, i.e. $V \in \mathcal{U}(F)$. From (I,j) we get $R \cap V = R \cap \bigcup_k (V \cap F_k) \subseteq W$, and the lemma is proved.

(2,11) PROPOSITION: *Let* R *be a sequentially closed subspace of* $F \in (LF)$. *Consider the conditions:*

(i) R *is well-located;*

(ii) $\iota_R \subseteq \tau(R, R'_\nu)$;

(iii) $\iota_R = \tau(R, R'_\nu)$;

(iv) *every sequentially continuous linear functional on* R *can be extended to a continuous linear functional on* F;

(v) R *is an* (LF)*-subspace;*

(vi) *every sequential neighborhood on* R *belongs to* $\mathcal{U}(R_\nu)$.

Then

$$(i) \Longleftrightarrow (ii) \Longleftrightarrow (iii) \Longleftrightarrow (iv) \Longleftarrow (v) \Longleftrightarrow (vi).$$

If F *is a Schwartz space, then all the above conditions are equivalent.*

PROOF: The implication (i) $\Longleftrightarrow$ (ii) is obvious. Since $R_\iota \in$ (LF), $\iota_R = \tau(R, R'_\iota)$. Hence (ii) $\Longleftrightarrow$ (iii). (iv) is a reformulation of (i) by means of the Hahn-Banach theorem. (v) $\Longleftrightarrow$ (vi) is trivial (cf. (2,9)). Finally, if F is a Schwartz space, then by (1,24), the mapping I in (1) is a homomorphism iff I is a weak homomorphism, i.e. (i) $\Longleftrightarrow$ (v).

(2,12) REMARK: Properties (ii), (iii) and (iv) are all characterizations of well-location in terms of duality. Intrinsinc characterizations of well-location were studied by W.Słowikowski to whom this concept is due (cf. [93, 94, 95, 96] and the bibliographical references in these articles)[1a]. Later we shall discuss a condition of nondual character, which is sufficient for (v) and thus also for (i) (cf. (2, 29) and (2,30)).

[1a] For other results on well-location, cf. [43, 76, 77, 83, 84].

(2,13) PROPOSITION: *Let* E *and* F *be* (LF)*-spaces and* $T \in L(E,F)$ *be a sequentially open mapping. Consider the following conditions:*

(a) T *is a homomorphism;*

(b) TE *is an* (LF)*-subspace;*

(c) T *is a weak homomorphism;*

(d) TE *is well-located in* F.

Then:

$$(a) \Longleftrightarrow (b) \Longrightarrow (c) \Longleftrightarrow (d)$$

If either E *or* F *is a Schwartz space, then all the above conditions are equivalent.*

PROOF: Set $R = TE$. The implication (a) $\Longrightarrow$ (b) is easy and the converse follows by the Dieudonné-Schwartz theorem (cf. (2,6)). Assume (c). Then $T: (E,\tau(E,R')) \longrightarrow (R,\tau(R,R'_\nu))$ is also a homomorphism. This combined with the sequential openness of T easily yields $\tau(R,R'_\nu) = \iota_R$, and (d) follows by (2,11). To prove the converse, we shall use Theorem (1,20). Let $v \in (T^{-1}(0))^o$. Let u_1 be the linear functional on R defined by $u_1(Tx) = \langle x,v \rangle$. The sequential openness of T implies that $u_1 \in R'_\iota$, i.e. $u_1 \in R'_\nu$. By the Hahn-Banach theorem we then get $u \in F'$ such that $T'u = v$. Thus we proved that $(T^{-1}(0))^o \subseteq T'F'$, which by (1,20) means that T is a weak homomorphism. If E is a Schwartz space, then (c) $\Longrightarrow$ (a) by (1,24). If F is a Schwartz space, then (d) $\Longrightarrow$ (b) by (2,11).

(2,14) The last proposition reduces problem (2,1) (cf. also the text following remark(2,6)) for sequentially open mappings to the study of (LF)-subspaces and well-located subspaces in F. Then it is natural to ask whether there exist (LF)-spaces containing a non-(LF)-subspace or a non-well-located subspace. The first example of an (LF)-space with a non-(LF)-subspace was found by Grothendieck in [32]. After the work of Hörmander [35] on strict P-convexity more natural examples of such spaces were found [42] [2]. In view of this we may ask: do there exist (LF)-spaces F such that every sequentially closed subspace of F is necessarily well-located? The answer to this question is given in (2,17). First we need two propositions due to B.M.Makarov [55]:

(2,15) PROPOSITION: *Let* F *be an (LF)-space such that*

(M) *every sequentially closed subspace of* F *is closed.*

Then every (sequentially) closed subspace of F *is also well-located.*

PROOF: Let $v \in R_\iota'$. Then $H = v^{-1}(0)$ is closed in R_ι, i.e., $H \cap F_n = H \cap R_n$ is closed in R_n and thus also in F_n. Hence by (M) H is closed in F, i.e. $v \in R_\nu'$.

(2,16) PROPOSITION: *Let* F *be a strict inductive limit of reflexive Banach spaces. Then* F *satisfies condition* (M) *of the last proposition.*

PROOF: Let F_j be the reflexive Banach spaces such that $F = \lim_j \text{ind}\, F_j$. Let S_j be the unit ball in F_j. We may assume that $S_1 \subseteq S_2 \subseteq \ldots$. Consider a sequentially closed subspace R of F and $x \in F \setminus R$. It suffices to show that $0 \notin \overline{R^*}$, where $R^* = -x + R$. This will follow if we can exhibit a sequence of $V_n \in \hat{\mathcal{U}}(F_n)$ such that

(3) $$V_n \subseteq V_{n+1}\,, \quad R^* \cap V_n = \emptyset \quad (n=1,2,\ldots).$$

Indeed, then for $V = \bigcup_{n=1}^{\infty} V_n$ we will have $V \in \mathcal{U}(F)$ and $V \cap R^* = \emptyset$. Set $R_j^* = R^* \cap F_j$ and choose $V_1 \in \hat{\mathcal{U}}(F_1)$ such that $V_1 \cap R_1^* = \emptyset$. F_1 being reflexive, V_1 is $\sigma(F_1, F_1')$-compact, hence also $\sigma(F_2, F_2')$-compact. Therefore in the topological space $\tilde{F}_2 = (F_2, \sigma(F_2, F_2'))$ there exist

2) Moreover, in this way we also obtain examples of closed non-well-located subspaces in nuclear (LF)-spaces [93, 34, 67, 84, 20].

non-intersecting neighborhoods of the closed set R_2^* and the compact set V_1, i.e. for some $V_2' \in \hat{\mathcal{U}}(\tilde{F}_2)$ we have $(V_1 + V_2') \cap (R_2^* + V_2') = \emptyset$. Set $V_2 = S_2 \cap (V_1 + V_2')$. Then $V_2 \in \hat{\mathcal{U}}(F_2)$, $V_1 \subseteq V_2 \subseteq S_2$, $V_2 \cap R^* = \emptyset$. Continuing in the same way we obtain (3).

(2,17) COROLLARY [55, 27, 12, 13]: *In strict inductive limits of reflexive Banach spaces all sequentially closed subspaces are closed and well-located.*

(2,18) Following (2,14), the assumption of normability of F_j's in the last statement cannot be dropped. Hence we may ask whether (2.15) holds when the spaces F_j are non-reflexive Banach spaces. It was shown by Pták [97] and Smoljanov [95] that it is not so: there exist (non-reflexive) (LB)-spaces containing closed but not well-located subspaces. (In view of their complicated nature we shall not discuss these examples here.) It is therefore interesting to look for a nontrivial characterization of well-location in (LB)-spaces:

(2,19) Let $F = \lim_j \text{ind}\, F_j$ be an (LB)-space and R a well-located subspace. Let S_n be the unit ball in F_n. We may assume that $S_1 \subseteq S_2 \subseteq \dots$. Let p_n be the seminorm [3)] in F' corresponding to the polar set $S_n^o \subseteq F'$. Hence

(4) $$p_1 \leqslant p_2 \leqslant \dots\,; \qquad S_n^o \supseteq F_n^o \quad (n = 1,2,\dots),$$

where F_n^o is a subspace of F'. Similarly, set $s_n = S_n \cap R_n$ and let q_n be the seminorm [3)] in R' corresponding to the polar $s_n^o \subseteq R'$. Then

---

3) By the inclusion in (4) ((5) resp.), p_n (q_n resp.) is not a norm.

(5) $$q_1 \leq q_2 \leq \ldots \quad ; \quad s_n^o \supseteq R_n^o \quad (n = 1,2,\ldots)$$

where R_n^o is a subspace of R'. The systems of seminorms $\{p_n\}$ and $\{q_n\}$ define the topology of the strong duals F_b' and R_ι' respectively. As strong duals of bornological spaces, both spaces F_b' and R_ι' must be complete hence Fréchet spaces. The adjoint to the continuous injection $J: R_\iota \longrightarrow F$ is a continuous mapping $J': F_b' \longrightarrow R_\iota'$ which must be surjective by (2). Furthermore, $J' = P(R)$ (cf. §0), for R_ι' and R_ν' coincide as vector spaces. Therefore, the Banach open mapping theorem applied to J' shows that for each n there is an integer m_n and a positive ε_n such that $P(R)S_n^o \supseteq \varepsilon_n s_m^o$ for all $m \geq m_n$ (cf. (5)). In particular, for every $\varepsilon > 0$, $P(R)S_n^o \supseteq \varepsilon R_{m_n}^o = R_{m_n}^o$ (cf. (5)). In terms of the seminorms the last inclusion says the following:

(P) There exists a sequence $m_n \nearrow \infty$ such that for any $\varepsilon > 0$ and any $u \in R_\iota'$ for which $P(R_{m_n})u = 0$ there exists $v \in F'$ such that $P(R)v = u$ and $p_n(v) < \varepsilon$.

Now assume that we start with $n = 1$. We can take $m_1 > 1$. Set $\ell_1 = 1$ and repeat the same construction with $n = m_1$. We get some $m_n > n$. Set $\ell_2 = m_n$. Next choose $n = m_1$, etc. In this way we get a sequence of integers $\ell_k \nearrow \infty$ such that if $H_n = F_{\ell_n}$, $\|\cdot\|_n = p_{\ell_n}$ and (H) denotes the statement, "$\exists (H_n, \|\cdot\|_n) \in (B) : F = \lim_n \text{ind}\, H_n$" [4], then the following holds:

(P_o) $$(H) : \forall n \in \mathbb{N} \quad \forall u \in (R \cap H_{n+1})^o \subset R_\iota' \quad \forall \varepsilon > 0 \quad \exists v \in F' : u = P(R)v \ \& \ \|P(H_n)v\|_n < \varepsilon.$$

This condition is obviously equivalent to the condition:

---

3) Hence $\|\cdot\|_n$ denotes both the norm in H_n and the seminorm in F' generated by this norm.

(P_1) $\quad$ (H): $\forall\, n \in \mathbb{N}\ \ \forall\, u \in (R \cap H_{n+1})^{\circ} \subset (R \cap H_{n+2})'$

$\forall\, \varepsilon > 0, \exists\, v \in H'_{n+2}\ :\ P(R \cap H_{n+2})v = u\ \&\ \|P(H_n)v\|_n < \varepsilon.$

Condition (P_1) is a special case of

(P_2) $\quad$ (H) $: \forall\, n \in \mathbb{N}\ \forall\, (u_1, u_2)\ :\ u_1 \in (H_n + R \cap H_{n+1})'\ \&$

$u_2 \in (R \cap H_{n+2})'\ \ \&\ P(R \cap H_{n+1})(u_1 - u_2) = 0$

$\forall\, \varepsilon > 0\ \exists\, w \in H'_{n+2}\ :\ P(R \cap H_{n+2})w = u_2\ \&\ \|P(H_n)w - P(H_n)u_1\|_n < \varepsilon.$

However it is easy to see that conversely $(P_1) \Longrightarrow (P_2)$: Indeed, fix u_1, u_2, ε as in (P_2) and let $\tilde{u}_i$ be an extension of u_i to H_{n+2} $(i = 1,2)$. Set $u = P(R \cap H_{n+2})(\tilde{u}_1 - \tilde{u}_2)$. Then u is as in (P_1). Hence there exists $v \in H'_{n+2}$ as in (P_1). Take $w = \tilde{u}_1 - v$. Then $w \in H'_{n+2}$ and $P(R \cap H_{n+2})w = P(R \cap H_{n+2})\left[(\tilde{u}_1 - \tilde{u}_2) + (\tilde{u}_2 - v)\right] = u + P(R \cap H_{n+2})(\tilde{u}_2 - v) = u + u_2 - P(R \cap H_{n+2})v = u_2$. Furthermore, $P(H_n)(w - u_1) = -P(H_n)v$ and (P_2) follows from (P_1).

Thus we have proved one part of the following statement:

(2,20) THEOREM [77] (cf. [83]): *Let* F *be an* (LB)-*space and* R *a sequentially closed subspace. Then* R *is well-located in* F *iff* R *satisfies one of the three equivalent conditions* (P_i) $(i = 0, 1, 2)$.

PROOF: It suffices to show that (P_1) implies the well-location of R. Assume that $\{H_n\}$ is a defining sequence with property (P_1) and set $R_n = R \cap H_n$. Let $r \in R'_l$ be arbitrary. Set $r_n = P(R_n)r$. Our aim is to find $q \in F'$ such that $P(R)q = r$. Let $q_2 \in F'$ be an arbitrary extension of r_2. Set $u_3 = r_3 - P(R_3)q_2$. Then $u_3 \in R_2^{\circ} \subset R'_3$ [5]; there exists, by (P_1), $q_3 \in F'$ such that $P(R_3)q_3 = u_3$ and $\|P(H_1)q_3\|_1 < \frac{1}{2}$. Consider $u_4 = r_4 - P(R_4)(q_2 + q_3)$. Since $u_4 \in R_3^{\circ} \subset R'_4$, there exists,

---

5) for the meaning of $R_i^{\circ} \subseteq R'_{i+1}$, cf. §0.

by (P_1), $q_4 \in F'$ such that $P(H_4)q_4 = u_4$ and $\|P(H_2)q_4\|_2 < \frac{1}{4}$. Continuing in this way we construct a sequence of functionals $q_j \in F'$ $(j = 2,3,\ldots)$ such that

(6)
$$r_j = P(R_j)(q_2 + \ldots + q_j) \quad (j = 3,4,\ldots)$$

and

(7)
$$\|P(H_{j-2})q_j\|_{j-2} < \frac{1}{2^{j-2}} \quad (j = 3,4,\ldots).$$

First we claim that $q = \sum_2^\infty q_j \in F'$. It suffices to show that the partial sums of q, restricted to any H_m, form a Cauchy sequence in the norm $\|\cdot\|_m$. Fix $m \geq 1$. Then by (7), for each $k \geq 2$,

$$\|P(H_m)q_{m+k}\|_m \leq \|P(H_{m+1})q_{m+1}\|_{m+1} \leq \ldots \leq \|P(H_{m+k-2})q_{m+k}\|_{m+k-2} \leq 2^{-m-k+2}.$$

This gives

$$\|P(H_m)\sum_{j=m+2}^{m+k} q_j\|_m \leq \frac{1}{2^{m-1}},$$

hence $q \in F'$. Next we claim that $P(R)q = r$, i.e. $P(R_n)q = r_n$ for every n. Indeed, for every $n \geq 1$, (6) gives

$$P(R_n)q_{n+1} = P(R_n)P(R_{n+1})q_{n+1} = P(R_n)\left[(r_{n+1} - P(R_{n+1})(q_2+\ldots+q_n)\right] =$$

$$= r_n - P(R_n)(q_2+\ldots+q_n) = P(R_n)q_n - P(R_n)q_n = 0.$$

Therefore also

$$P(R_n)q_{n+k} = P(R_n)P(R_{n+k-1})q_{n+k} = 0$$

for every $n \geqslant 1$ and $k \geqslant 1$. This together with (6) gives

$$P(R_n)q = P(R_n)(q_2+\ldots+q_n) = r_n ,$$

which proves the theorem.

(2,21) REMARK: Despite their seemingly technical character, both the theorem and its proof are relatively transparent: If R is well-located in F, then the Banach open mapping theorem applied to the dual spaces F' and R_1' yields immediately condition (P_o), of which conditions (P_1) and (P_2) are only simple reformulations. The proof of

$$\text{"}(P_1) \Longrightarrow \text{well-location of } R\text{"}$$

consists of an inductive construction of a functional $q \in F'$. This is done by extending q to larger and larger spaces H_n. The construction is carried out so that: (i) the "correction" added at each step vanishes on the space H_{n-1} considered in the previous step. At the same time this correction "corrects" the previously obtained functional so that it extends r_n to F'; (ii) the whole process converges. This proof suggests two remarks: First, it resembles the proof of the classical Mittag-Leffler theorem about the decomposition of a meromorphic function into partial fractions. Second, it looks similar to a proof of the vanishing some of cohomology. Actually, both remarks are related: In [66] Palamodov obtained an abstract homological version of the Mittag-Leffler theorem which represents a general basis for various results, some of which are related to (2,20) (cf. also [83]).

Let us return to (2,14). The simplest example of a nuclear (LF)-space which contains a closed but not well-located subspace is the space $\mathcal{D}' = \mathcal{D}'(\mathbb{R}^1)$ of Schwartz distributions on the real line[6]. This follows from the work of Hörmander [35] on convolution operators (cf. [93, 84, 67]). It was observed in [81, 34] that such spaces lead easily to the examples of spaces which are complete but not fully complete. More generally, we have the following:

(2,22) PROPOSITION [81]: *Let* F *be a reflexive* (LF)*-space which does not have property* (M) *of* (2,15). *Then the strong dual* F' *is complete but not fully complete.*

PROOF: Let $F = \lim_n \text{ind}\, F_n$. Since F' is obviously complete, it suffices by (1,5) to exhibit a subspace H of $F'' = F$, which is almost closed but not closed. By our hypothesis, F contains a subspace H which is sequentially closed but not closed. We shall prove that H is almost closed. Let $\mathcal{B}(F)$ be the family of all bounded sets in F. Take any $U \in \hat{\mathcal{U}}(F')$. Because of the reflexivity of F we may assume that $U^o = A^{oo}$ for some $A \subseteq \mathcal{B}(F)$. Since A^{oo} is bounded, $A^{oo} \subseteq F_{n_o}$ for some $n_o \in \mathbb{N}$. Since A^{oo} is closed and absolutely convex,

$$U^o \cap H = A^{oo} \cap H$$

is $\sigma(F,F')$-closed. Hence H is almost closed.

(2,23) COROLLARY: *The space* $\mathcal{D}'$ *is not fully complete. More generally, if* F *is a reflexive* (LF)*-space cointaining a closed but not well-located subspace, the strong dual* F' *is complete but not fully complete.*

REMARK: Actually, one can prove that $\mathcal{D}'$ is not even hereditary complete (cf. [34, 67]).

---

6) and more generally, $\mathcal{D}'(\Omega)$ for a suitable open subset $\Omega \in \mathbb{R}^n$. Since these spaces are necessarily Schwartz spaces, we cannot obtain in this way a well-located subspace which would not be an (LF)-subspace (cf. (2,11)).

In the rest of this chapter we shall discuss a result of Pták [75] which gives a sufficient condition for a subspace R of $F \in (LF)$ to be an (LF)-subspace.

(2,24) LEMMA [76]: *Let* F *be a l.c. space and* X *and* Y *two subspaces of* F. *Then the following two conditions are equivalent:*

(a) The canonical mapping $Q: X \oplus Y \longrightarrow X + Y$ *is weakly open.*

(b) Given $x' \in X'$ *and* $y' \in Y'$ *which coincide on* $X \cap Y$, *there exists a simultaneous extension* $z' \in F'$ *of both* x' *and* y'.

PROOF: By (1,20) it suffices to show that (b) is equivalent to

(a'): $Q'(X + Y)' \supseteq (Q^{-1}(0))^{o}$.

Let $z' \in F'$. In view of $Q'F' \subseteq X' \oplus Y'$, we have $Q'z' = (x', y')$ for some $x' \in X'$ and $y' \in Y'$. Then for each $w \in X \oplus Y$, $w = (x, y)$,

$$\text{(8)} \qquad \langle x,x'\rangle + \langle y,y'\rangle = \langle (x,y), (x',y')\rangle = \langle w, Q'z'\rangle$$

$$= \langle Qw, z'\rangle = \langle x+y, z'\rangle.$$

In particular, (8) implies that $x' = P(X)z'$ and $y' = P(Y)z'$, i.e., $Q'(z') = (P(X)z', P(Y)z')$ for all $z' \in F'$. Since obviously

$$Q^{-1}(0) = \{(t, -t) : t \in X \cap Y\},$$

we get

$$Q^{-1}(0)^{o} = \{(x', y') \in X' \oplus Y' : P(X \cap Y)x' = P(X \cap Y)y'\}.$$

Hence $(a') \Longleftrightarrow (b)$.

(2,25) DEFINITION: Let X and Y be two closed subspaces of a Fréchet space F. Then X and Y are called mutually *orthogonal* (notation $X \perp Y$) in F, if one of the following equivalent conditions holds:

1º The canonical mapping $Q: X \oplus Y \longrightarrow X + Y$ is a homomorphism.

2º The mapping Q of 1º is a weak homomorphism.

3º Condition (b) of (2,24).

REMARKS: The equivalence of 1º, 2º and 3º follows from (1,21) and (2,24). – $X \perp Y$ (in F) iff $Y \perp X$ (in F). – Let X, Y and F be as in (2,25), and let F be a subspace of a Fréchet space G. Then $X \perp Y$ (in F) iff $X \perp Y$ (in G).

(2,26) DEFINITION: Let R be a sequentially closed subspace of an (LF)-space F. We say that R is *orthogonal* in F, if there exists a defining sequence $F_j \longrightarrow F$ such that $F_j \perp R \cap F_k$ for each $k > j \geq 1$.

A dual characterization of this concept is given in the next proposition. Since the proof is straightforward but rather technical (cf. [75]), it will be omitted. On the other hand, this proposition will not be used in the sequel.

(2,27) PROPOSITION: *Let R be a sequentially closed subspace of an (LF)-space F. Then R is orthogonal in F iff*

(p) for every defining sequence F_j of F there exists a sequence of integers $p(j) \geq j$ $(j = 1,2,\ldots)$ such that given $k > p(j)$, $y' \in F'_{p(j)}$ and $r' \in (R \cap F_k)'$ such that

$$P(R \cap F_{p(j)})(y' - r') = 0,$$

there exists $x' \in F'_k$ such that

$$P(F_j)(x' - y') = 0 \quad \text{and} \quad P(R \cap F_k)x' = r'.$$

(2,28) LEMMA: *Let X be a vector space with Y and Z subspaces. Furthermore, assume that L, M and N are absolutely convex subsets of X such that $N \subset Y$, $N \cap Z \subset M$ and $L \cap (Y+Z) \subset N + (M \cap Z)$. Fix an arbitrary $\sigma \in (0, \frac{1}{2})$ and let H be the convex hull of $\sigma L \cup (1 - \sigma)N$. Then $H \cap Z \subset M$.*

The proof is straightforward.

(2,29) THEOREM [75]: *Let* F *be an* (LF)-*space and* R *a sequentially closed orthogonal subspace of* F. *Then* R *is an* (LF)-*subspace.*

PROOF: Let $F = \lim_j \mathrm{ind}\, F_j$ and $R_k = R \cap F_k$. We may assume that the spaces F_j are chosen so that $F_j \perp R_k$ for all $k > j \geq 1$. Fix an arbitrary sequential neighborhood W in R. Let $j \geq 1$ and $\varepsilon \in (0,1)$ be fixed. Let $V_j \in \mathcal{U}(F)$ satisfy (I,j) of (2,4). By (2,25) the mapping $Q: F_j \oplus R_{j+1} \longrightarrow F_j + R_{j+1}$ is a homomorphism. Hence there is $\tilde{W} \in \hat{\mathcal{U}}(F)$ such that

$$\tilde{W} \cap (F_j + R_{j+1}) \subseteq (V_j \cap F_j) + (V_{j+1} \cap R_{j+1}).$$

Setting in (2,28), $X = F$, $Y = F_j$, $Z = R_{j+1}$, $L = \tilde{W}$, $M = W$, $N = V_j \cap F_j$ and $\sigma = \frac{\varepsilon}{2}$ we get, for $H = \mathrm{cv}((1-\sigma)(V_j \cap F_j) \cup \sigma\tilde{W})$, $H \cap R_{j+1} \subseteq U$. In view of

$$0 < \mu = \frac{1-\varepsilon}{1-\sigma} < 1,$$

the last inclusion shows that $V^*_{j+1} \overset{\mathrm{def.}}{=} \mu H$ satisfies condition (I,j+1). Moreover, $V^*_{j+1} \in \mathcal{U}(F)$, since $V^*_{j+1} \supset \mu\sigma W$. On the other hand, by the definition of H,

$$(1-\varepsilon)(V_j \cap F_j) = \mu(1-\sigma)[V_j \cap F_j] \subseteq \mu H = V^*_{j+1}.$$

If we set $V_{j+1} = V^*_{j+1} \cap V_j$, we see that condition (II,j) of (2,10) is satisfied and the theorem follows.

(2,30) COROLLARY [75]: *Let* E *and* F *be two* (LF)-*spaces and* $T \in L(E,F)$ *be a sequentially open mapping such that its range* TE *is orthogonal in* F. *Then* T *is a homomorphism.*

Proof follows from (2,29) and (2,13).

REMARKS: A remark similar to (2,21) could be made about (2,29) and (2,30). Furthermore, it is interesting to compare condition (P_2) with condition (p) of (2,27). On the basis of this comparison it is natural to call with Pták a sequentially closed subspace R of F *semiorthogonal* in F, if it satisfies one of the equivalent conditions (P_i) of (2,19). Theorem (2,29) can be proved (in a more general form) by methods of homological algebra (cf. [84]).

§ 3 - FREE LOCALLY CONVEX SPACES AND RELATED TOPICS

(3,1) Consider a category $\mathcal{L}$ and a subcategory $\mathcal{K}$. Given $X \in \mathcal{L}$, then an object X' in $\mathcal{K}$ is called a *reflection* of X (in the subcategory $\mathcal{K}$) if there is an isomorphism of functors $\mathrm{Hom}_{\mathcal{K}}(X',\cdot): \mathcal{K} \longrightarrow \mathcal{E}ns$ and $\mathrm{Hom}_{\mathcal{L}}(X,\cdot): \mathcal{K} \longrightarrow \mathcal{E}ns$. In other words, X' is a reflection of X if the functor $\mathrm{Hom}_{\mathcal{L}}(X,\cdot): \mathcal{K} \longrightarrow \mathcal{E}ns$ can be represented by X' (cf. [28,67]). It is easy to see that this is the case iff there exists a morphism $\omega_X \in \mathrm{Hom}(X,X')$ such that the above mentioned functorial isomorphism

$$\Phi: \mathrm{Hom}_{\mathcal{K}}(X',\cdot) \longrightarrow \mathrm{Hom}_{\mathcal{L}}(X,\cdot)$$

is of the form $\Phi = \omega^*$ where $\omega^*(u) = u\omega_X$ for any morphism $u \in \mathrm{Hom}_{\mathcal{L}}(X',Z)$ an arbitrary object Z of $\mathcal{K}$. Furthermore, if X_1' is another reflection of X, then $X_1' \cong X'$ +). If a reflection X' exists for every $X \in \mathcal{L}$, then the mapping $X \longmapsto X'$ defines a functor Λ from $\mathcal{L}$ into $\mathcal{K}$, and $\mathcal{K}$ is said to be a *reflective subcategory* of $\mathcal{L}$. Similarly one defines *coreflection* and *coreflective* subcategories (For further information on these topics, cf. [8, 28, 67]).

In this chapter we shall study the case when $\mathcal{L}$ is the category *TCR* of all completely regular spaces and $\mathcal{K}$ is the subcategory *TLC*. Let us recall that by a completely regular space X we mean a space

---

+) Hence we shall speak about *the* reflection of X (cf. (3,5)).

satisfying the axiom (U) of Urysohn (cf. [4]) and no further separation axiom. C(X) denotes the space of all continuous functions $f: X \longrightarrow \mathbb{K}$. It will be shown below that *TLC* is a reflective subcategory of the category *TCR*. The existence of the reflection $\Lambda: TCR \longrightarrow TLC$ is easy to establish (cf. (3,6)). On the other hand, the reflections $\Lambda(X)$ of $X \in TCR$ can be described intrinsically and have interesting properties (cf. (3,8), (3,9), etc.). Furthermore, we shall briefly discuss some other concepts arising naturally in connection with Λ.

(3,2) To begin with, let us assume that a given $X \in TCR$ has the reflection $\Lambda(X)$ and $\omega_X: X \longrightarrow \Lambda(X)$ is the canonical morphism. This means that for any continuous mapping $v: X \longrightarrow F$, where $F \in TLC$ is arbitrary, there exists a unique $w \in L(\Lambda(X), F)$ such that $v = w\omega_X$, i.e. the diagram

(1)
$$\begin{array}{ccc} X & & \\ {\scriptstyle \omega_X}\downarrow & \searrow^{v} & \\ \Lambda(X) & \xrightarrow[w]{} & F \end{array}$$

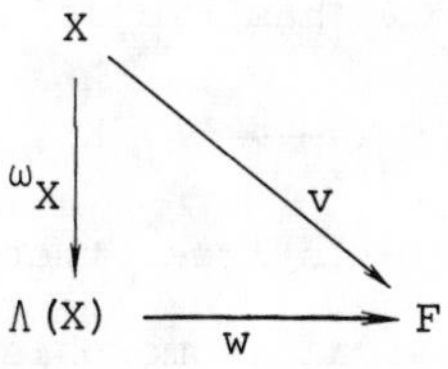

is commutative. We shall write $w = \lambda(v)$ and call $\lambda(v)$ the *linearization* of v.

(3,3) LEMMA: *Let* $X \in TCR$ *have the reflection* $\omega_X: X \longrightarrow \Lambda(X)$ *in the category* TLC. *Then*

(a) *the range* $\omega_X(X)$ *of* ω_X *is a total subset of* $\Lambda(X)$;

(b) ω_X *is an injective mapping;*

(c) *if* $X \in TLC$, *then* $\lambda(1_X): \Lambda(X) \longrightarrow X$ *is a surjective homomorphism of ℓ.c. spaces.*

PROOF: (a) follows from the Hahn-Banach theorem. To prove (b), let F be the free module of X over $\mathbb{K}$, F endowed with the weakest ℓ.c. topology. If $v: X \longrightarrow F$ is the canonical injection, it must be continuous. Hence we conclude from (1) that ω_X is injective. Finally, (c) follows from the continuity of ω_X.

(3,4) DEFINITION: Given $X \in TCR$, the reflection $\Lambda(X)$ is called the *free ℓ.c. space of* X, provided that

(α) $\omega_X(X)$ is a Hamel basis of the vector space $\Lambda(X)$;

and,

(β) ω_X is a homeomorphic embedding of X into $\Lambda(X)$.

(3,5) Since reflections are determined up to an isomorphism in *TLC*, it follows that if a space $X \in TCR$ has two reflections

$$\omega_X^i: X \longrightarrow \Lambda_i(X), \quad i = 1,2,$$

and one of them – say, $\Lambda_1(X)$ – is free, then $\Lambda_2(X)$ must also be free. Moreover, there exists a canonical isomorphism $\iota: \Lambda_1(X) \longrightarrow \Lambda_2(X)$ such that $\iota \circ \omega_X^1 = \omega_X^2$. (To see this, it suffices to apply (1) to $F = \Lambda_2(X)$ with $\Lambda = \Lambda_1$, and then to reverse the roles of $\Lambda_i(X)$.) Since we usually identify x with $\omega_X^i(x)$, we can say that ι leaves X fixed.

(3,6) PROPOSITION: *The free ℓ.c. space* $\Lambda(X)$ *exists for each* $X = (X,t) \in TCR$, *and is determined uniquely up to isomorphisms which leave* X *fixed. Moreover,* $\omega_X: X \longrightarrow \Lambda(X)$ *has the following properties: Let* $\mathcal{T}_X$ *be the topology of the free ℓ.c. space* $\Lambda(X)$ *and let*

$\mathbf{T}_1$ ($\mathbf{T}_2$ *resp.*) *be the family of all l.c. topologies* $\mathcal{T}$ *on the vector space* $\Lambda(X)$ *such that* $\omega_X\colon (X,t) \longrightarrow (\Lambda(X), \mathcal{T})$ *is a continuous (homeomorphic resp.) injection. Then*

$$\mathcal{T}_X = \sup \mathbf{T}_1 = \sup \mathbf{T}_2. \tag{2}$$

PROOF: The uniqueness of $\Lambda(X)$ being already established in (3,5), we shall construct the space $(\Lambda(X), \mathcal{T}_X)$ on the basis of properties it should have according to the theorem and (3,3). The vector space $\Lambda(X)$ can be defined as the free module of X over $\mathbb{K}$, i.e. as the set of all formal linear combinations $\sum_1^m \mu_i x_i$ $(\mu_i \in \mathbb{K},\ x_i \in X)$. Let

$$\omega_X\colon X \longrightarrow \Lambda(X)$$

be the corresponding injection of X into $\Lambda(X)$, i.e. $\omega_X(x) = 1.x$. Then every mapping v of the set X into any vector space X extends uniquely to a linear mapping $\lambda(v)\colon \Lambda(X) \longrightarrow F$. With t, $\mathbf{T}_1$ and $\mathbf{T}_2$ being defined as in the statement of the proposition, set

$$\tilde{\Lambda}(X) = \{\lambda(f) : f \in C(X)\}.$$

The complete regularity of X implies that

$$\sigma = \sigma(\Lambda(X), \tilde{\Lambda}(X)) \in \mathbf{T}_2.$$

Hence $\emptyset \neq \mathbf{T}_2 \subseteq \mathbf{T}_1$. Set $\mathcal{T}_X = \sup \mathbf{T}_1$. Then $t = \omega_X^{-1}(\sigma) \subseteq \omega_X^{-1}(\mathcal{T}_X)$. The continuity of $\omega_X\colon (X,t) \longrightarrow (\Lambda(X), \mathcal{T}_X)$ gives $t \supseteq \omega_X^{-1}(\mathcal{T}_X)$. Therefore, $\mathcal{T}_X \in \mathbf{T}_2$ and thus also (2) follows. It remains to show that $(\Lambda(X), \mathcal{T}_X)$ has the property (1). With $F = (F, \tilde{\mathcal{T}})$ and v as in (1), set $w = \lambda(v)$ and $\mathcal{T}^* = w^{-1}(\mathcal{T})$. The continuity of v shows that

$$\omega_X^{-1}(\mathcal{T}^*) = w^{-1}(\tilde{\mathcal{T}}) \subseteq t,$$

i.e. $\mathcal{T}^* \in \mathbf{T}_1$, which gives $\mathcal{T}^* \subseteq \mathcal{T}_X$. The continuity of w is thus established.

Actually this proof also yields the following:

(3,7) COROLLARY: *Set* $\tilde{\Lambda}(X) = \{\lambda(f) : f \in C(X)\}$ *and* $\sigma = \sigma(\Lambda(X), \tilde{\Lambda}(X))$. *Then every* $\mathcal{T} \in \mathbf{T}_1$ *such that* $\mathcal{T} \supseteq \sigma$ *we have* $\mathcal{T} \in \mathbf{T}_2$ *and*

$$(\Lambda(X), \mathcal{T})' = \tilde{\Lambda}(X).$$

(3,8) THEOREM [78]: *Let* $X \in TCR$. *Denote by* $\mathcal{H}$ *the family of all subsets* H *of* C(X) *such that*

(*i*) *for every* $x \in X$, $\sup_{f \in H} |f(x)| < \infty$, *i.e. the set* $\lambda(H)$ *is weakly bounded in* $\tilde{\Lambda}(X)$;

(*ii*) *for every* $x \in X$ *and* $\varepsilon > 0$, *the set*

$$W_{H,\varepsilon}(x) = \{y \in X : |f(x) - f(y)| < \varepsilon, f \in H\}$$

is a neighborhood of x *in* X, *i.e.* H *is an equicontinuous subset of* C(X).

Then the topology $\mathcal{T}_X$ *of the free l.c. space* $\Lambda(X)$ *can be defined as the topology* $\mathcal{T}(\mathcal{H})$ *of uniform convergence on all sets of the form* $\lambda(H)$, $H \in \mathcal{H}$.

PROOF: It follows by definition that $\sigma \subseteq \mathcal{T}(\mathcal{H}) \in \mathbf{T}_1$, hence $\mathcal{T}(\mathcal{H}) \subseteq \mathcal{T}_X$, and by (3,7), $\mathcal{T}(\mathcal{H}) \in \mathbf{T}_2$. Thus it suffices to show that $\mathcal{T}(\mathcal{H}) \supseteq \mathcal{T}_X$. By (2) this will follow if we prove that $\mathcal{T}(\mathcal{H}) \supseteq \mathcal{T}$ for any $\mathcal{T} \in \mathbf{T}_1$, $\mathcal{T} \supseteq \sigma$. Fix such a $\mathcal{T}$ and denote by $\mathcal{P}$ the system $Spec(\Lambda(X), \mathcal{T})$. Furthermore, set $V_p = \{z \in \Lambda(X) : p(z) \leqslant 1\}$ for each $p \in \mathcal{P}$. Let V_p^o be the polar of V_p in $\tilde{\Lambda}(X)$, and $H_p = \{g\omega_X : g \in V_p^o\}$. Every V_p is a barrel, hence H_p satisfies condition (i) of the theorem. Moreover,

$$\begin{aligned} H_p &= \{f \in C(X) : |\langle z, \lambda(f)\rangle| \leqslant p(z), z \in \Lambda(X)\} \\ &\subseteq \{f \in C(X) : |f(x) - f(y)| \leqslant p(1\cdot x - 1\cdot y)\}, \end{aligned}$$

where as usual, $1\cdot x = \omega_X(x)$, etc. For $\varepsilon > 0$ and $x \in X$ fixed, define

$$N(x) = \{y \in X : p(1\cdot x - 1\cdot y) < \varepsilon\}.$$

$N(x)$ is a neighborhood of x in X, because $\mathcal{T} \in \mathbf{T}_1$. Thus, if $y \in N(x)$ and $f \in H_p$, then $|f(x) - f(y)| < \varepsilon$, i.e. $N(x) \subseteq W_{H_p,\varepsilon}(x)$. This shows that the set H_p satisfies condition (ii). Therefore, $\mathcal{T} \subseteq \mathcal{T}(\mathcal{H})$ which proves the theorem.

(3,9) COROLLARY: *The following properties are equivalent:*

(i) X *is a Hausdorff completely regular space;*

(ii) $(\Lambda(X), \mathcal{T}_X)$ *is a Hausdorff l.c. space;*

(iii) $\omega_X(X)$ *is a closed subset of* $(\Lambda(X), \mathcal{T}_X)$.

PROOF: If X is Hausdorff, $C(X)$ distinguishes points in X, whence the topology σ is Hausdorff. This shows that (i) $\Longleftrightarrow$ (ii). If X is Hausdorff and $u \in \Lambda(X) \setminus \omega_X(X)$, then it is easy to find a neighborhood W of u such that $W \cap \omega_X(X) = \emptyset$. If X is not Hausdorff, there must exist points $x, y \in X$ such that $f(x) = f(y)$ for all $f \in C(X)$. But then every neighborhood $W_{H,\varepsilon}(z)$ of the point

$$z = \frac{1}{2}x + \frac{1}{2}x \notin \omega_X(X)$$

contains points $1\cdot x$ and $1\cdot y$ from the set $\omega_X(X)$. Hence $\omega_X(X)$ cannot be closed.

Next we shall investigate how certain properties of a completely regular space X are "reflected" in the properties of the space $\Lambda(X)$. (Unless stated otherwise, $\Lambda(X)$ will always mean the free ℓ.c. space of X.)

(3,10) PROPOSITION: *(a) Let* $X = \bigoplus_{\iota \in I} X_\iota$ *be the direct sum of completely regular spaces* X_ι, $\iota \in I$. *Then*

$$(3) \qquad \Lambda(X) = \bigoplus_{\iota \in I} \Lambda(X_\iota).$$

(b) If Z *is either a finite or an isolated subset* [1] *of* X *and* $n = \mathrm{card}(Z)$, *then*

$$(4) \qquad \Lambda(X) = \mathbb{K}^{(n)} \oplus \Lambda(X \setminus Z).$$

PROOF: (a) is obvious. Hence (b) follows when Z is an isolated subset of X. If Z is finite, then $\Lambda(Z) \cong \mathbb{K}^{(n)}$ as a finite dimensional subspace of $\Lambda(X)$, must have a topological complement in $\Lambda(X)$.

(3,11) COROLLARY: *Let* X_ι, $\iota \in I$, *be completely regular spaces and* Z *a subspace of* $X = \bigoplus_{\iota \in I} X$ *such that for every* $\iota \in I$

$$Z \cap X_\iota = Z_\iota^1 \cup Z_\iota^2,$$

where Z_ι^1 *is finite and* Z_ι^2 *is an isolated subset of* X_ι, *and* $Z_\iota^1 \cap Z_\iota^2 = \emptyset$. *Then relation* (4) *again holds.*

Let X and Y be completely regular spaces and $f: X \longrightarrow Y$ a continuous mapping. By the general properties of reflections (cf. [28, 67]), there is a unique mapping $\Lambda(f) \in L(\Lambda(X), \Lambda(Y))$ such that the diagram

$$(5) \qquad \begin{array}{ccc} X & \xrightarrow{\ f\ } & Y \\ {\scriptstyle \omega_X}\downarrow & & \downarrow{\scriptstyle \omega_Y} \\ \Lambda(X) & \xrightarrow[\Lambda(f)]{} & \Lambda(Y) \end{array}$$

---

1) i.e. a set each point of which is isolated in X.

commutes. It is natural to ask under what conditions on f, the mapping $\Lambda(f)$ is a surjective homomorphism.

(3,12) Let us recall the following terminology: Given topological spaces X and Y and a surjective mapping $f: X \longrightarrow Y$, then Y is called a *quotient space* of X and f a *quotient mapping*, if

(Q) G is open in Y iff $f^{-1}(G)$ is open in X.

Obviously, every continuous open mapping is a quotient mapping. Although the converse is in general false (cf. (3,16)), in some special cases it does hold: Let $f: X \longrightarrow Y$ be a continuous group homomorphism of topological groups X and Y. Then f is open iff f is a quotient mapping. (The proof is obvious.)

(3,13) PROPOSITION: *Let* $f: X \longrightarrow Y$ *be a quotient mapping,* X *and* Y *completely regular spaces. Then* $\Lambda(f): \Lambda(X) \longrightarrow \Lambda(Y)$ *is a surjective homomorphism of the free l.c. spaces.*

PROOF: Set $F = \Lambda(f)$. If $\mathcal{T}_X$ and $\mathcal{T}_Y$ denote as usual the topologies of the free l.c. spaces $\Lambda(X)$ and $\Lambda(Y)$, define the topology $\mathcal{T}^*$ on $\Lambda(Y)$ as the F-image of the topology $\mathcal{T}_X$ (F is clearly surjective!). The proposition asserts that $\mathcal{T}^* = \mathcal{T}_Y$. In view of the continuity of F we only have to show that $\mathcal{T}^* \subseteq \mathcal{T}_Y$. By (2), it suffices to establish the continuity of $\omega_Y: Y \longrightarrow (\Lambda(Y), \mathcal{T}^*)$. Let U be open in $(\Lambda(Y), \mathcal{T}^*)$ and $U \cap \omega_Y(Y) \neq \emptyset$. The set $H = \omega_X^{-1}(F^{-1}(U))$ is obviously open in X. But $H = f^{-1}(\omega_Y^{-1}(U))$. Hence by (Q), $\omega_Y^{-1}(U)$ must be open in Y. The proposition then follows.

The following example is due to V.Geĭler [30] and will be used in the next chapter:

(3,14) EXAMPLE: We shall use the following terminology: Let $\mathfrak{m}$ be a fixed infinite cardinal. Whenever A and B are sets such that $A \subseteq B$ and $\mathrm{card}(A) \leqslant \mathfrak{m} < \mathrm{card}(B)$, we shall say that A is a *hole* in B.

Choose arbitrarily a set M of cardinality $>\mathfrak{m}$, a topological space X, and an arbitrary point t in X. Then a new topological space X_t can be defined as follows:

The set X_t consists of the pair (t,t) and all pairs of the form (x,m) where $x \in X \setminus \{t\}$ and $m \in M$. Let $p_t\colon X_t \longrightarrow X$ be the projection of each pair on the first coordinate. The topology in X_t is defined by the following two conditions: (i) the neighborhoods of the point (t,t) are all sets of the form

$$W = p_t^{-1}(V) \setminus \bigcup \{A_x : x \in X \setminus \{t\}\}, \tag{6}$$

where V is any neighborhood of t in X, and for each $x \in X \setminus \{t\}$, A_x is an arbitrary hole in $\{x\} \times M$; (ii) all other points of X_t are isolated.

It follows immediately from these definitions (and, in particular, from the possibility of "making holes into the neighborhoods of (t,t)") that:

(A) X_t is a Hausdorff completely regular space (regardless of whether X was Hausdorff or not).

(B) Each hole in X_t is a discrete subspace of X.

(C) Every open convering of X_t can be refined to an open decomposition of X_t.[2)]

(D) $p_t\colon X_t \longrightarrow X$ is continuous but not a quotient mapping.

---

2) Spaces with this property are sometimes called perfectly zero-dimensional.

Moreover, if X is not discrete, p_t does not admit any continuous section.

On the other hand, the mapping p_t behaves at the "singular point" (t,t) as a quotient mapping. This indicates that if one could find a new space T and a projection $p: T \longrightarrow X$ such that for each $x \in X$, the set $p^{-1}(x)$ would contain a singularity of type (t,t), then p would probably be a quotient mapping. The point is to define the space T so that T would still have properties (A), (B) and (C) of the space X_t:

(3,15) PROPOSITION: *If* X, M *and* $\mathfrak{m}$ *are as above, define the space* T *by*

$$T = T(X, M, \mathfrak{m}) = \bigoplus_{t \in X} X_t.$$

Then T *has properties* (A), (B) *and* (C) *of* X_t *, and moreover,*

(D') $\quad p = \bigoplus_{t \in X} p_t : T \longrightarrow X$ *is a quotient mapping.*

PROOF: Properties (A), (B) and (C) are obvious. The points of T can be viewed as triples of either of two kinds: (1) (t,t,t) for $t \in X$; (2) (t,x,m) where $t \neq x$ are arbitrary points in X and $m \in M$. The mapping p is then the projection of each triple on its second coordinate. Let G be a subset of X such that $p^{-1}(G)$ is open in T. Choose an arbitrary point τ in G. Then, for each $t \in X$, the set $p_t^{-1}(G) = p^{-1}(G) \cap X_t$ is open in X_t. In particular, $p_\tau^{-1}(G)$ contains a set W of the form (6) where V is now a neighborhood of the point τ in X. Therefore, $G = p_\tau(p_\tau^{-1}(G)) \supset p_\tau(W) = V$. Hence G is open in X and this proves the proposition.

(3,16) REMARK: It is obvious that for a suitable X, the quotient mapping p define in (D') will not be open.

Our interest in the space T stems from the following property of T:

(3,17) PROPOSITION: *Let* $T = T(X,M,\mathfrak{m})$ *be as above. Then the free* ℓ.c. *space* $\Lambda(T)$ *of* T *is such that any subspace* E *of* $\Lambda(T)$ *with* $\mathrm{card}(E) \leqslant \mathfrak{m}$ *is necessarily isomorphic to the space* $\mathbb{K}^{(\mathfrak{n})}$ *for some cardinal* $\mathfrak{n} \leqslant \mathfrak{m}$.

PROOF: Let Z be the set of all $z \in T$ such that $\omega_T(z)$ occurs in at least one expression of the form $\sum \mu_i \omega_T(z_i) \in E$. Then

$$\mathfrak{p} = \mathrm{card}(Z) \leqslant \mathfrak{m}.$$

Furthermore, Z is a discrete subspace of T (cf. (3,15)(B)), and satisfies the hypotheses of (3,11) for $X = T$. Hence $\Lambda(Z) \cong \mathbb{K}^{(\mathfrak{p})}$. But then E as a subspace of $\Lambda(Z)$ must be isomorphic to a subspace of $\mathbb{K}^{(\mathfrak{p})}$, i.e. $E \cong \mathbb{K}^{(\mathfrak{n})}$ for some $\mathfrak{n} \leqslant \mathfrak{m}$.

We can summarize the preceding sections as follows: The category *TLC* is a reflective subcategory of the category $\mathcal{L}$ = *TCR*. Moreover, for each $X \in$ *TCR*, the reflection $\Lambda(X)$ is actually the free ℓ.c. space associated with X. $\Lambda(X)$ can be simply described (cf. (3,6),(3,8)) and enjoys some interesting properties (cf. (3,3)-(3,5),(3,9)-(3,13)).

However, there is another topological category $\mathcal{L}$ which also contains *TLC* as a subcategory, and in some sense is even more closely related to the category *TLC* than the category *TCR*, namely the category $\mathcal{L}$ = *UN* of uniform spaces. Indeed, it suffices to recall that each space $E \in$ *TLC* carries the natural uniformity which is the only uniformity compatible with the linear structure of E. Hence one can ask about the reflectivity of TLC in the category UN. The answer is again in the affirmative, and all the results (an their proofs) stated above for $\mathcal{L} = TLC$ remain valid – after minor modifications – also in the case $\mathcal{L} = UN$ [3]. Let us only quote one consequence of the reflectivity of TLC in UN:

(3,18) PROPOSITION: *Every uniform space* X *can be embedded by a uniform homeomorphism into a ℓ.c. space* E *so that the image of* X *is closed in* E *iff* X *is a Hausdorff uniform space.*

Actually, the problem of embedding metrizable (resp. uniform) spaces into normal (resp. ℓ.c.) spaces was posed many years ago. It has been studied by several authors independently of the concept of a free ℓ.c. space (cf. the references in [1] and also in the review of [1] in Math. Rev.,18 (1957), p.406). Later a simple proof of (3,18) was given in [63] (for more information about the relationship between (3,18) and the free ℓ.c. spaces, see the last section of [78].) All these authors considered the problem of embedding a topological structure into a linear space from the point of view of general topology rather than functional analysis. In particular, properties of spaces dual to linear extensions were not considered.

The definition of the free ℓ.c. space of a completely regular space was given by A.A.Markov in 1941 [58]. Markov was led to this concept by his investigations of free topological groups where a similar

---

3) thus,e.g., in condition (ii) of (3,7) one has to take H uniformly equicontinuous, etc.

situation occurs (cf. also [59]). In [58] Markov formulates property (3,2) and asserts that for each $X \in \mathcal{TCR}$, the *l.c.* space $\Lambda(X)$ exists and is unique in the sense of (3,5). (No proofs were given in [58]. In 1964 D.A.Raĭkov [78] published a proof of (3,6) and found also (3,8) and (3,9).)

The fact that linear extensions enable us to consider *linear* combinations of points in *non-linear* topological spaces suggests possible connections of this concept to measure theory on the one hand (cf."combinaisons des masses ponctuelles" in [6]; cf. also [40]), and functional analysis on the other hand (cf. Stone-Čech compactifications, duality between $C(X)$ and $C(X)'$, etc.)[4]. Some of these will be discussed in the rest of this chapter.

Henceforth only Hausdorff spaces X will be considered, and $\Lambda(X)$ will always mean the *vector* space $\Lambda(X)$. According to (3,7), the dual of the free *l.c.* space $(\Lambda(X), \mathcal{T}_X)$ is the space $\tilde{\Lambda}(X)$ of linearizations $\lambda(f)$ of all continuous functions f on X. In addition to the topologies $\mathcal{T}_X$ and $\sigma_X = \sigma(\Lambda(X), \tilde{\Lambda}(X))$, each X determines uniquely the Mackey-Arens topology $\tau_X = \tau(\Lambda(X), \tilde{\Lambda}(X))$. Hence $\sigma_X \subseteq \mathcal{T}_X \subseteq \tau_X$ and the interval $\sigma_X \subseteq \mathcal{T} \subseteq \mathcal{T}_X$ contains exactly all those *l.* c. topologies $\mathcal{T}$ on $\Lambda(X)$, for which the natural embedding $\omega_X: X \longrightarrow (\Lambda(X), \mathcal{T})$ is homeomorphic (or, equivalently, continuous). It will be useful to simplify our notation as follows: $C(X)$ will be identified with $\tilde{\Lambda}(X)$. The duality between $\Lambda(X)$ and $C(X)$ is then given by the formula

---

4) Actually, it follows from results of Pták [72,74] that both aspects of linear extensions are related; moreover, in applications, arguments based on duality between linear extensions and their duals can often replace arguments using integration.

(7) $$\langle x,f\rangle = \sum_{i=1}^{m} \mu_i f(x_i)$$

where $x = \sum_{i=1}^{m} \mu_i \omega_X(x_i) \in \Lambda(X)$ and $f \in C(X)$. This identification of $\tilde{\Lambda}(X)$ with $C(X)$ enable us to consider $\Lambda(X)$ as a linear subspace of $C(X)'$. (Nevertheless, this notation can also be misleading. Consider, for instance, a $\ell.c.$ topology $\mathcal{T}$ on $\Lambda(X)$ such that $\sigma_X \subseteq \mathcal{T} \subseteq \tau_X$. Hence $(\Lambda(X), \mathcal{T})' = C(X)$. What is then an equicontinuous subset of $C(X)$? Clearly, this can mean two different things: First, H can be an equicontinuous subset in the sense of general topology, i.e. H as in condition (ii) of (3,8). Second, it can mean an equicontinuous subset in the dual space to $(\Lambda(X), \mathcal{T})$, i.e. $H \subseteq U^o$ where $U \in \mathcal{U}(\Lambda(X), \mathcal{T})$. Raĭkov's theorem (3,8) says that in the case of $\mathcal{T} = \mathcal{T}_X$ these two meanings are equivalent. More precisely, if $H \subseteq C(X)$ is of the form U^o with U as above, then there exists an equicontinuous set (in the first sense) $H_1 \subseteq C(X)$ such that $H \subseteq H_1^{oo}$, where the polars are always taken with respect to the bilinear form (7). However, if $\mathcal{T} \neq \mathcal{T}_X$, then this is no longer true in view of the same theorem.)

The first duality result related to the linear extensions of topological spaces was obtained by Pták [70] in 1954:

(3,19) THEOREM: *Let* X *be a pseudocompact completely regular space and* B *an absolutely convex subset of* $C(X)$. *Then* B *is* $\sigma(C(X), C(X)')$-*compact if (and only if)* B *is* $\sigma(C(X), \Lambda(X))$-*compact.*

(For the proof, cf. [70].)

However, a systematic study of ℓ.c. spaces of the form $(\Lambda(X), \mathcal{T})$ and their duals was initiated by M.Katětov [44, 45, 46] who considered the following more general problem.

Let X be a completely regular (or uniform) space and $\mathcal{W}$ a linear subspace of $C(X)$ such that the bilinear form (7), considered on the product $\Lambda(X) \times \mathcal{W}$, is nondegenerate in both variables. Therefore, the vector spaces $\Lambda(X)$ and $\mathcal{W}$ form a dual pair. Describe the completion of $\Lambda(X)$ equipped with the topology $\tau(\Lambda(X), \mathcal{W})$. Similarly, if $\mathcal{W} = C(X)$, describe the completion of the free ℓ.c. space $(\Lambda(X), \mathcal{T}_X)$.

Katětov obtained several interesting results of this type, and these were later reproved and extended in various ways (cf. [38, 39, 100, 102]). We shall discuss only the simplest case when X is compact and $\mathcal{W} = C(X)$.

(3,20) THEOREM: *Let* X *be a compact space. Then there are isomorphisms of the vector spaces,*

(8) $$(\Lambda(X), \mathcal{T}_X)^{\wedge} \cong (\Lambda(X), \tau_X)^{\wedge} \cong C(X)'.$$

PROOF: In view of (1,2) it suffices to show that

(9) $$(\Lambda(X), \mathcal{T}_X)^{\wedge} \subseteq C(X)' \subseteq (\Lambda(X), \tau_X)^{\wedge}.$$

To prove the first inclusion in (9), take any $\theta \in (\Lambda(X), \mathcal{T}_X)^{\wedge}$. By (1,1), θ is $\sigma(C(X), \Lambda(X))$-continuous on every equicontinuous subset of $(\Lambda(X), \mathcal{T}_X)'$. In particular, consider a sequence $f_n \in C(X)$, $f_n \longrightarrow 0$ uniformly on X, and set $H = \{f_n\} \cup \{0\}$. By (3,8), there exists a set H_1 which is equicontinuous in $(\Lambda(X), \mathcal{T}_X)'$ and such that $H \subseteq H_1^{oo}$. Since the sequence $\{f_n\}$ also converges weakly in $C(X)$, it converges in the topology $\sigma(C(X), \Lambda(X))$ too. But θ being $\sigma(C(X), \Lambda(X))$-continuous on H_1^{oo}, this gives $\theta(f_n) \longrightarrow 0$, i.e. $\theta \in C(X)'$. – The second inclusion in (9) will follow, if we can show that every $\theta \in C(X)'$ is $\sigma(C(X), \Lambda(X))$-continuous on all equicontinuous subsets U^o of $(\Lambda(X), \tau_X)'$. But this is an immediate consequence of (3,19).

The preceding theorem can be generalized in various ways. Thus, for instance, if X is a precompact uniform space, then $(\Lambda(X), \mathcal{T}_X)\hat{}$ and $C(X)'$ are linearly isomorphic. Similarly, for X pseudocompact and completely regular, there is a linear isomorphism

$$(\Lambda(X), \mathcal{T}_X)\hat{} \cong C(\beta X)'$$

(cf. [39, 74, 100, 103, 104]).

The above choice of X and $\mathcal{W}$ is not the only interesting one. If we take, e.g. $X = [0, 1]$ and $\mathcal{W} = C^\infty[0, 1]$, then $(\Lambda(X), \mathcal{T}_X)\hat{}$ can be identified with the space $\mathcal{E}'(X)$ of all distributions with support in $[0, 1]$ (cf. [39, 102]). If suitably modified, this type of linear extension yields the distribution spaces $\mathcal{D}'(X)$ and $\mathcal{E}'(X)$ on an arbitrary smooth variety X as different completions of the linear extension $\Lambda(X)$ of this variety [38]. Moreover, in this way we can even obtain some natural concepts of differential geometry on varieties. There are other important function spaces which can be obtained by suitable linear extensions [18]. On the other hand, linear extensions of topological spaces turn out to be very useful in the study of weak compactness in ℓ.c. spaces, completions of uniform spaces, etc. For these and other results on linear extensions, the reader is referred to [26, 57, 62, 101, 103, 104].

§ 4 - INJECTIVE AND PROJECTIVE LOCALLY CONVEX SPACES

In this chapter we shall discuss certain properties of the category *TLC*, which are essential for applications of homological methods to the study of ℓ.c. spaces.

(4,1) Throughout the whole chapter by a morphism we shall always mean a morphism in the category *TLC*. Furthermore, let us recall that monomorphisms (epimorphisms) in *TLC* are exactly all injective (surjective) morphisms. The surjectivity of epimorphisms easily follows from the fact that among objects of the category *TLC* there are vector spaces endowed with the weakest ℓ.c. topology (i.e. a topology in which the only neighborhood of the origin is the whole space). For the sake of brevity we shall often say that a given morphism in mono (epi, iso, homo, resp.), if f is a monomorphism (epimorphism,..., resp.). If f is both mono (epi resp.) and homo, then f will be called *monohomo* (*epihomo* resp.). Obviously, $f: X \longrightarrow Y$ is monohomo iff f is an isomorphic embedding of X onto a subspace of Y. Similarly, $f: X \longrightarrow Y$ is epihomo iff f defines a quotient mapping of X onto the quotient space $X/f^{-1}(0) \cong Y$ (cf. (3,12)).

More generally, let $f: X \longrightarrow Y$ be any morphism. Then f can be factorized as follows

(1)
$$\begin{array}{ccc} X/f^{-1}(0) & \xrightarrow{\tilde{f}} & f(X) \\ \uparrow{\scriptstyle \lambda_f} & & \downarrow{\scriptstyle \mu_f} \\ X & \xrightarrow[f]{} & Y \end{array}$$

where λ_f is the natural quotient mapping and μ_f is the natural embedding. Hence, for every f, λ_f is epihomo and μ_f is monohomo. Moreover, $\tilde{f}$ is always both mono and epi; but $\tilde{f}$ is iso iff f is homo. The fact that $\tilde{f}$ is not always iso is the only reason why *TLC* fails to be an Abelian category (cf. [8, 28]). This shows that we cannot apply the standard Abelian homology to the category *TLC* without certain modifications:

(4,2) DEFINITION: Let I and P be given ℓ.c. spaces. Then I is called an *injective space* (or simply: I is injective), if every morphism $f: X \longrightarrow I$, defined on a subspace X of an arbitrary ℓ.c. space Y, can be extended to a morphism $F: Y \longrightarrow I$. Similarly, P is said to be a *projective space* (or: P is projective), if every morphism $g: P \longrightarrow Y/X$ can be lifted to a morphism $G: P \longrightarrow Y$.

In diagrams these properties look as follows:

(2a)

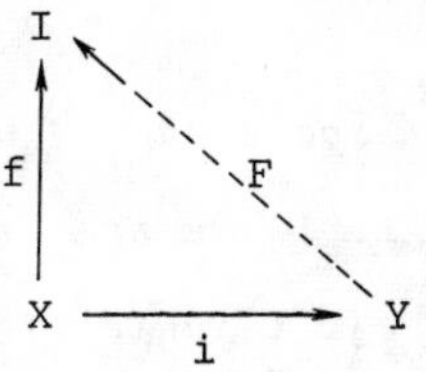

(i monohomo, f morphism)

(2b)

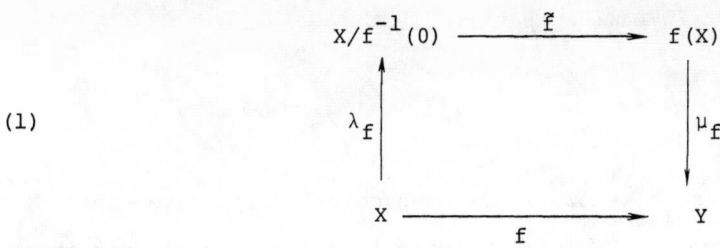

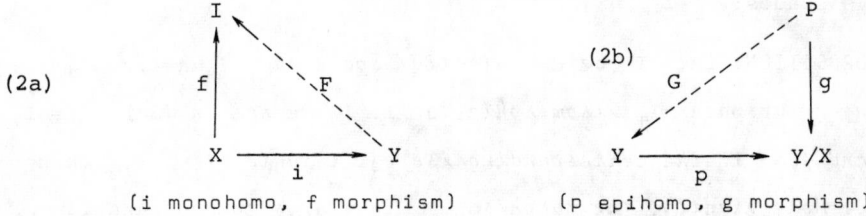

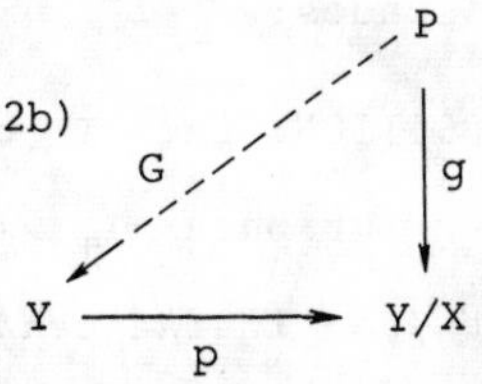

(p epihomo, g morphism)

Now it is natural to pose the following problems:

(I) *Characterize all injective spaces.*

And similarly,

(P) *Characterize all projective spaces.*

For our purposes, however, it is even more important to know whether every $\ell.c.$ space X is isomorphic to a subspace of some injective space; in other words,

(i) *does the category TLC possess sufficiently many injectives?*

And similarly, is every $\ell.c.$ space X isomorphic to a quotient space of some projective space, i.e.

(p) *does the category TLC possess sufficiently many projectives?*

While (I) represents a classical problem of functional analysis in TVS (cf. [41, 51, 52, 65]), (i) is not difficult to answer in the affirmative (cf. [67] and (4,19) below). On the other hand, it was shown recently by V.A.Geĭler [30] that the problem (P) has a very simple solution (cf. (4,7)) which yields immediately a negative answer to the question (p) (cf. (4,8)). In the sequel we shall discuss these topics in detail. As an application, we shall obtain an interesting result of V.P.Palamodov ([67]; cf. also (4,22)).

To begin with, let us state some simple properties of injective and projective spaces:

(4,3) PROPOSITION: *Let* I *be an injective space and* Y *a* $\ell.c.$ *space containing a subspace* I_1 *isomorphic to* I. *If we set in* (2a), $X = I$, $f = 1_I$ *and for* i *the corresponding iso of* I *onto* $I_1 \subseteq Y$, *then* $Y = I_1 \oplus Y_1$. *More exactly, the morphism* F *in* (2a) *is epihomo and is uniquely determined by the canonical projection* $I_1 \oplus Y_1 \longrightarrow I_1$. *Similarly, let* P *be a projective space and* $X \subseteq Y$ *l.c. spaces such that there is an iso* $g: P \longrightarrow Y/X$. *Then the mapping* G *in* (2b) *is uniquely determined;* G *is monohomo and* $Y = G(P) \oplus X$.

PROOF: It suffices to apply Propositions 13 and 14 of [5], Chap. I, §1.

(4,4) PROPOSITION: *Let* $\{I_\alpha, \alpha \in A\}$ *be an arbitrary family of injective spaces. Then their cartesian product is also injective. Analogously, the direct sum* $\bigoplus_{\alpha \in I} P_\alpha$ *of an arbitrary family of projectives is also projective.*

PROOF follows easily from definitions.

(4,5) PROPOSITION: *The field* $\mathbb{K}$ *considered as a l.c. space over itself is both injective and projective.*

PROOF: The injectivity of $\mathbb{K}$ is equivalent to the Hahn-Banach theorem. The rest is easy.

(4,6) COROLLARY: *For any cardinal* $\mathfrak{n}$, $\mathbb{K}^{\mathfrak{n}}$ *is injective and* $\mathbb{K}^{(\mathfrak{n})}$ *is projective.*

(4,7) THEOREM [30]: P *is a projective space iff* $P = \mathbb{K}^{(\mathfrak{n})}$ *for some cardinal* $\mathfrak{n}$.

PROOF: In view of (4,6) we may assume that P is projective. According to (3,3)(c), $\lambda(1_P): \Lambda(P) \longrightarrow P$ is an epihomo of the free l.c. space $\Lambda(P)$ onto P. Set $\mathfrak{m} = \text{card}(P)$ and fix an arbitrary set of $\text{card}(M) > \mathfrak{m}$. Consider the space $T = T(P, M, \mathfrak{m})$ and the quotient mapping $p: T \longrightarrow P$ defined in (3,15)(D'). Then by (3,13) the morphism $\Lambda(p): \Lambda(T) \longrightarrow \Lambda(P)$ is epihomo so that

(3)

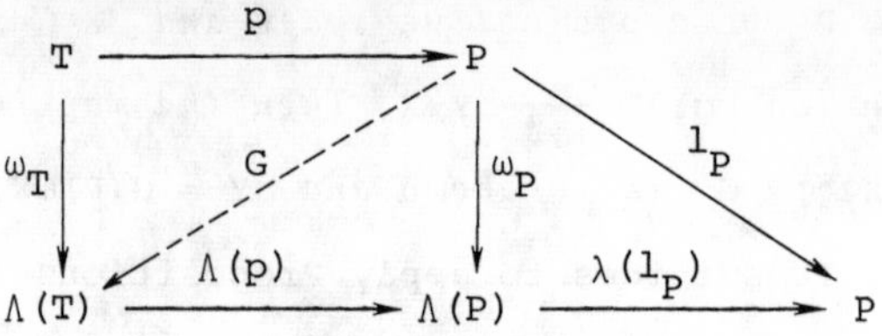

$g = \lambda(1_P) \circ \Lambda(p)$ is also epihomo. Since P is projective, we obtain $G: P \longrightarrow \Lambda(T)$ such that $g \circ G = 1_P$. By (4,3), G defines an embedding of P onto a direct summand of $\Lambda(T)$. The theorem then follows from (3,17).

(4,8) COROLLARY [30]: *The category TLC does not have sufficiently many projectives.*

PROOF: Indeed, if it did, then, for instance, every ℓ.c. space would be Hausdorff and bornological.

Although the category *TLC* is a subcategory of the category of all vector spaces over $\mathbb{K}$, the category V can also be embedded in two natural ways as a full subcategory of *TLC*:

(4,9) DEFINITION: Given $X \in V$, let $\mathfrak{s}(X)$ ($\mathcal{S}(X)$ resp.) be the weakest (finest resp.) ℓ.c. topology on X. Let ${}^{\mathfrak{s}}V$ be the subcategory of *TLC*, whose objects are all ℓ.c. spaces ${}^{\mathfrak{s}}X = (X, \mathfrak{s}(X))$ and

$$\mathrm{Hom}_{{}^{\mathfrak{s}}V}({}^{\mathfrak{s}}X, {}^{\mathfrak{s}}Y) = L({}^{\mathfrak{s}}X, {}^{\mathfrak{s}}Y).$$

Similarly, ${}^{\mathcal{S}}V$ is defined as the subcategory of *TLC*, whose objects are all spaces ${}^{\mathcal{S}}X = (X, \mathcal{S}(X))$ and

$$\mathrm{Hom}_{{}^{\mathcal{S}}V}({}^{\mathcal{S}}X, {}^{\mathcal{S}}Y) = L({}^{\mathcal{S}}X, {}^{\mathcal{S}}Y).$$

(4,10) PROPOSITION: *If for any* $X \in TLC$, $|X|$ *denotes the underlying vector space, then for arbitrary* $X, Y \in TLC$, $Y = {}^{s}Y$,

(4) $$\mathrm{Hom}_{TLC}(X, {}^{s}Y) = \mathrm{Hom}_{{}^{s}V}({}^{s}X, {}^{s}Y) = \mathrm{Hom}_{V}(|X|\ ,\ |Y|).$$

Similarly, for arbitrary X, Y TLC, $X = {}^{\mathcal{S}}X$, *we have*

(5) $$\mathrm{Hom}_{TLC}({}^{\mathcal{S}}X,\ Y) = \mathrm{Hom}_{{}^{\mathcal{S}}V}({}^{\mathcal{S}}X, {}^{\mathcal{S}}Y) = \mathrm{Hom}_{V}(|X|\ ,\ |Y|).$$

Proof is obvious.

(4,11) COROLLARY: *Both* ${}^{s}V$ *and* ${}^{\mathcal{S}}V$ *are full subcategories of TLC, and are isomorphic to* V.

(4,12) PROPOSITION: *Let* ${}^{\mathcal{S}}X \in {}^{\mathcal{S}}V$ *and* $\mathfrak{n}$ *be the cardinality of a Hamel basis of* ${}^{\mathcal{S}}X$. *Then* ${}^{\mathcal{S}}X \cong \mathbb{K}^{(\mathfrak{n})}$. *In particular, every* ${}^{\mathcal{S}}X \in {}^{\mathcal{S}}V$ *is projective.*

PROOF: Let $(e_\iota)_{\iota \in I}$ be a Hamel basis of ${}^{\mathcal{S}}X$ and $\mathfrak{n} = \mathrm{card}(I)$. Then it is easy to see that ${}^{\mathcal{S}}X = \bigoplus_{\iota \in I} \mathbb{K}e_\iota \cong \mathbb{K}^{(\mathfrak{n})}$ (cf. [5], Chap. II, §2, nº 4, example 2). By (4,6), ${}^{\mathcal{S}}X$ is projective.

Similarly, we have:

(4,13) PROPOSITION: *Every space* ${}^{s}W \in {}^{s}V$ *is injective.*

PROOF: Let $f: X \longrightarrow Y$ be monohomo and $v: X \longrightarrow {}^{s}W$ a morphism. By the first equation in (4), the identity mappings $i: X \longrightarrow {}^{s}X$ and $i': Y \longrightarrow {}^{s}Y$ define the reflections of X and Y in ${}^{s}V$. Hence there is a unique ${}^{s}f: {}^{s}X \longrightarrow {}^{s}Y$ such that the square in the diagram

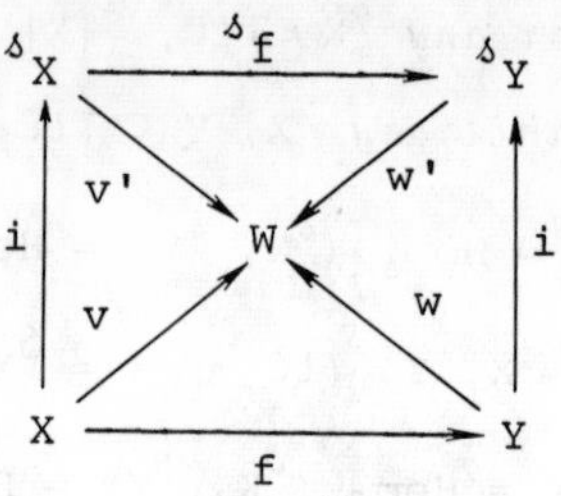

is commutative. Furthermore, there is a unique $v'\colon {}^{s}X \longrightarrow {}^{s}W$ such that $v = v'i$. Next we claim that there is a $w'\colon {}^{s}Y \longrightarrow {}^{s}W$ such that $v' = w'\,{}^{s}f$. In view of the second equation in (4), it suffices to construct an arbitrary linear mapping with this property. If Y_1 is an arbitrary algebraic complement of ${}^{s}f({}^{s}X)$ in ${}^{s}Y$, then by (4,11) we also have ${}^{s}Y = {}^{s}f({}^{s}X) \oplus Y_1$ in the category *TLC*. Thus w' can be defined by setting $w' \equiv 0$ on Y_1 and $w' = v'({}^{s}f)^{-1}$ on ${}^{s}f({}^{s}X)$. Set $w = w'i'$. Then $wf = w'i'f = w'\,{}^{s}fi = v'i = v$, which proves the proposition.

(4,14) PROPOSITION: *${}^{s}V$ is both a reflective and coreflective subcategory of the category TLC. If $X \in TLC$, the reflection of X in ${}^{s}V$ in defined by $i\colon X \longrightarrow {}^{s}X$ where i is the identity mapping. The coreflection of X is defined by $j\colon X_s \longrightarrow X$ where X_s denotes the closure of the origin in X, i.e. X_s is the maximal subspace H of X on which the induced topology from X equals s(H); j is the natural embedding.*

PROOF: The reflectivity of ${}^{s}V$ follows from (4). That $j\colon X_s \longrightarrow X$ defines a coreflection of X in ${}^{s}V$ follows easily from (4,15) and (4,3).

(4,15) THEOREM (D.A.Raĭkov; cf. [67]): *$\mathcal{S}_V$ is a coreflective but not reflective subcategory of TLC.*

PROOF: The coreflectivity of $\mathcal{S}_V$ follows from (5). Hence it remains to exhibit a ℓ.c. space X which does not have a reflection in $\mathcal{S}_V$. Set $X_n = \mathbb{K}$, $n \in \mathbb{N}$, and $X = \prod_{n=1}^{\infty} X_n = \mathbb{K}^{\mathbb{N}}$. Let $p_n: X \longrightarrow X_n$ be the corresponding projections. Assume that X has a reflection in $\mathcal{S}_V$ given by $p: X \longrightarrow X^+$. Then for any $\mathcal{S}_Z \in \mathcal{S}_V$, the mapping $f \longmapsto fp$ defines a bijection

(6) $$\mathrm{Hom}_{\mathcal{S}_V}(X^+, \mathcal{S}_Z) \longrightarrow \mathrm{Hom}_{TLC}(X, \mathcal{S}_Z)$$

which shows that p must be epi. Setting in (6), $\mathcal{S}_Z = X_n$, we obtain for each $n \in \mathbb{N}$, the unique morphism $q_n: X^+ \longrightarrow X_n$ such that $q_n p = p_n$:

(7)

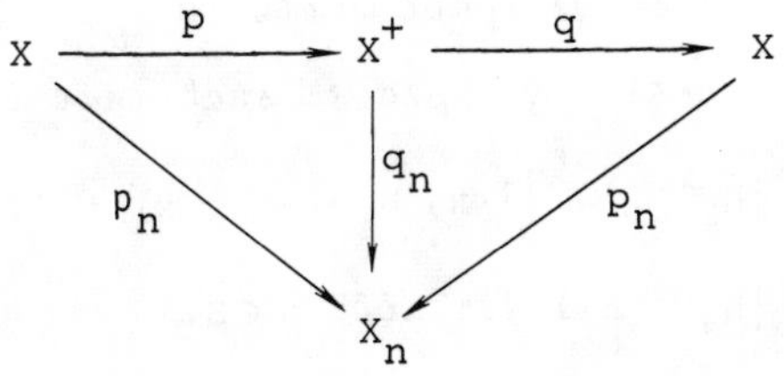

Since X is a cartesian product, there is a unique $q: X^+ \longrightarrow X$ such that $q_n = p_n \circ q$ for all $n \in \mathbb{N}$. Therefore the diagram (7) is commutative. In particular, $p_n = p_n \circ q \circ p$, $n \in \mathbb{N}$, which yields $q \circ p = 1_X$ and thus also $p \circ q \circ p = p$. Since p is epi, the last identity gives $p \circ q = 1_{X^+}$. Hence $X \cong X^+$ which shows that the topology of X is $\mathcal{S}(X)$. But this is a contradiction.

We have seen in (4,6) that for an arbitrary non empty set S, the space $\mathbb{K}^S$ is injective. $\mathbb{K}^S$ can be viewed as the space of all functions $f: S \longrightarrow \mathbb{K}$ with the topology of uniform convergence on finite subsets of S. Let $m(S)$ be the subspace of $\mathbb{K}^S$ consisting of all bounded functions on S. The topology of $m(S)$ is defined by the uniform convergence on S. Hence $m(S)$ is a Banach space with the usual sup-norm. We shall show that $m(S)$ is also an injective space. First we need a simple lemma whose proof is trivial:

(4,16) LEMMA: *Consider a l.c. space* E *and a subset*

$$B = \{h_s \in E' \ , \ s \in S\}$$

of the dual space E'. *Then the following properties are equivalent:*

(a) B *is equicontinuous.*

(b) *There is a unique* $\psi: E \longrightarrow m(S)$ *such that*

$$\langle x \ , \ h_s \rangle_X = [\psi(x)](s).$$

In this case we shall write $B = B_\psi$.

(c) *There exists* $q \in \mathit{Spec}\ E$ *such that if*

$$\|h_s\|_q \overset{\text{def}}{=} \{|\langle x, h_s\rangle| \ : \ q(x) \leqslant 1, \quad x \in E\},$$

then $\|h_s\|_q \leqslant 1$ *for all* $s \in S$.

(4,17) PROPOSITION: *For every* $S \neq \emptyset$, *the space* $m(S)$ *is injective.*

PROOF: Let $X \subseteq Y$ and $\psi: X \longrightarrow m(S)$. By (4,16) there is a $q \in \mathit{Spec}\ X$ such that $\|h_s\|_q \leqslant 1$ for all $h_s \in B_\psi$. Since X is a subspace of Y, we can assume that $q \in \mathit{Spec}\ Y$. Extending each $h_s \in B_\psi$ to Y so that for the extension h_s^* we have $\|h_s^*\|_q = \|h_s\|_q$, we obtain an equicontinuous set B^* in Y'. Hence by (4,16), $B^* = B^*_{\psi^*}$ for some $\psi^*: Y \longrightarrow m(S)$ which obviously extends ψ.

REMARK: The last proposition is just a different formulation of the Hahn-Banach theorem. For a deep result of this kind, cf. the Nachbin-Kelley theorem and related facts (cf. [41, 48, 52, 64, 65]).

(4,18) PROPOSITION: *Given* $X \in TLC$, *let* $j: X_s \longrightarrow X$ *be the reflection of* X *in* V (cf. (4,14)). *If* T *is any basis of* $\mathcal{U}(X)$, *set*

$$I(X) = X_s \times \prod_{U \in T} m(U^o)$$

where as usual U^o *is the polar of* X *in* X'. *Then there exists a monohomo* $\theta_X: X \longrightarrow I(X)$.

PROOF: For each $U \in T$ let X_U be the seminormed space $(X, \|\cdot\|_U)$, where $\|\cdot\|_U$ denotes the seminorm defined by U. Let $\theta_U(x) \in m(U^o)$ be the function $s \longmapsto \langle x,s \rangle$. If $\tilde{U}$ is the unit ball in $m(U^o)$, then $\theta_U^{-1}(\tilde{U}) = U$. Whence (i) $\theta_U: X_U \longrightarrow m(U^o)$ is homo; and,

(ii) $$\theta_U \in L(X, m(U^o)).$$

Set $\tilde{X} = \prod_{U \in T} m(U^o)$. By (ii) we can form the morphism

$$\theta_o = \prod_{U \in T} \theta_U: X \longrightarrow \tilde{X}$$

and by (i) θ_o is homo. Moreover $\theta_o^{-1}(0) = \bigcap_{U \in T} U = X_s$, and thus by (1), $\tilde{\theta}_o: X/X_s \longrightarrow \tilde{X}_s$ is monohomo. According to (4,3), X_s has a topological complement X_1 in X, hence $X_1 \cong X/X_s$. Then it suffices to define $\theta_X = 1_{X_s} \oplus \theta_o: X \longrightarrow X_s \oplus \tilde{X} \cong I(X)$.

REMARK: As the reader has certainly observed, the last proposition is a simple modification of the well-known fact that every Hausdorff ℓ.c. space can be embedded into a product of Banach spaces (cf. [5], Chap. II, §5, Prop. 7). Nevertheless, Proposition (4,18) has important consequences:

(4,19) COROLLARY [67]: *The category TLC has sufficiently many injectives.*

PROOF: Indeed, this follows from (4,4), (4,13), (4,17) and (4,18).

(4,20) COROLLARY: *Every ℓ.c. space* X *has an exact injective resolution,*

(8) $$0 \longrightarrow X \xrightarrow{\varepsilon} I_0 \xrightarrow{i_0} I_1 \xrightarrow{i_1} I_2 \xrightarrow{i_2} \cdots \ .$$

In other words, there exist injective spaces I_k $(k \geq 0)$ *and homomorphisms* ε *and* i_k $(k \geq 0)$ *such that the sequence* (8) *is exact, i.e.* ε *is monohomo,* $\varepsilon(X) = i_0^{-1}(0)$ *and* $i_k(I_k) = i_{k+1}^{-1}(0)$ *for all* $k \geq 0$. *In particular,* $X \cong i_0^{-1}(0)$.

PROOF: Define $I_0 = I(X)$, $\varepsilon = \theta_X$ as in (4,18). If p_0 is the quotient mapping $p_0: I_0 \longrightarrow X_0 = I_0/\varepsilon(X)$, set $I_1 = I(X_0)$ and $i_0 = \theta_X \circ p_0$. Obviously, i_0 is homo and $i_0^{-1}(0) = p_0^{-1}(0) = \varepsilon(X)$. Similarly, we get $i_1: I_1 \longrightarrow I_2$, etc.

Corollary (4,20) represents the starting point for applications of homological algebra to the category *TLC*. Indeed, if $F: TLC \longrightarrow A$ is some functor of *TLC* into an Abelian category A, we can define the right derivatives R^qF of the functor F and apply the powerful techniques of homological algebra to the study of F. On the other hand, many of the important problems of functional analysis in ℓ.c. spaces can be formulated in terms of such functors. We shall conclude this chapter by demonstrating this fact on a simple example.

Consider again the main problem of functional analysis in TVS: Given $f \in L(X, Y)$, decide whether f is a homomorphism or not. This problem is obviously of topological character. Nevertheless, V.P.Palamodov showed that it can be reduced to a purely algebraic problem (cf. [66, 67]). First we shall start with a special case of his result, formulated in terms of standard functional analysis:

(4,21) PROPOSITION [67]: *Let* $f \in L(X, Y)$. *Then* f *is a homomorphism if and only if*

($\mathcal{P}$) *every equicontinuous set* $B = \{h_s, s \in S\}$ *such that* $B \subseteq (f^{-1}(0))^{o}$, *can be lifted by means of* f *to an equicontinuous subset*

$$B^* = \{h_s^*, s \in S\}$$

in Y', *i.e.*

$$<f(x), h_s^*>_Y = <x, h_s>_X$$

for all $x \in X$ *and* $s \in S$.

Before proving this proposition let us establish a simple lemma:

LEMMA: *The following conditions are equivalent to condition* ($\mathcal{P}$):

($\mathcal{P}$, $m(S)$): *For every* $S \neq \emptyset$ *and arbitrary* $\psi: X \longrightarrow m(S)$ *such that* $\psi \in f^{-1}(0)$ *there exists* $\chi: Y \longrightarrow m(S)$ *such that* $\psi = \chi f$.

($\mathcal{P}$, $I(Z)$): *The same condition as* ($\mathcal{P}$, $m(S)$), *but with* $m(S)$ *replaced by the space* $I(Z)$ (cf. (4,18)) *constructed for an arbitrary l.c. space* Z.

PROOF OF THE LEMMA: By (4,16), ($\mathcal{P}$)$\Longleftrightarrow$($\mathcal{P}$, $m(S)$). Since ($\mathcal{P}$, $I(Z)$) trivially implies ($\mathcal{P}$, $m(S)$), it suffices to show the converse. In view of the fact that $I(Z)$ is defined as a cartesian product, it suffices to prove ($\mathcal{P}$, $I(Z)$) with $I(Z)$ replaced by any factor of this product. Hence it remains to prove $(\mathcal{P}, Z_\delta)$ for any $Z \in \mathcal{TLC}$. But this is trivial, because if $\psi: X \longrightarrow Z_\delta$ is any morphism such that $\psi = 0$ on $f^{-1}(0)$, then by (4), we can take χ any linear mapping $\chi: Y \longrightarrow Z_\delta$ (cf. (4,10)) such that $\psi = \chi f$.

PROOF OF (4,21): Let f be a homomorphism and let B be any set as in $(\mathcal{P})$. By (4,16), $B = B_\psi$ for some $\psi: X \longrightarrow \mathfrak{m}(S)$ and $\psi = 0$ on $f^{-1}(0)$. Hence there is $\psi': X/f^{-1}(0) \longrightarrow \mathfrak{m}(S)$ such that $\psi = \psi'\lambda_f$.

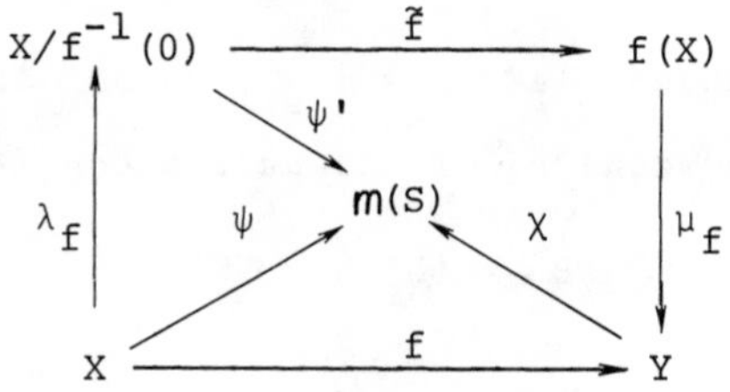

Since f is homo, $\tilde{f}$ is iso and therefore $\psi = \psi'\lambda_f$ is monohomo. Then the injectivity of $\mathfrak{m}(S)$ (cf. (4,17)) yields $\chi: Y \longrightarrow \mathfrak{m}(S)$ such that $\psi' = \chi\mu_f\tilde{f}$. Hence

$$\psi = \psi'\lambda_f = \chi\mu_f\tilde{f}\lambda_f = \chi f.$$

If $B^* = \{h_s^*, s \in S\}$ is the equicontinuous set in Y' defined by means of χ (cf. (4,16)), i.e.

$$\langle y, h_s \rangle_Y = [\chi(y)](s)$$

for all $y \in Y$ and $s \in S$, then for $x \in X$ we have

$$\langle f(x), h_s \rangle_Y = [\chi(f(x))](s) = [\psi(x)](s) = \langle x, h_s \rangle_X ,$$

which proves condition $(\mathcal{P})$.

Conversely, assume that $(\mathcal{P})$ – and thus also $(\mathcal{P}, I(Z))$ – holds. Consider the diagram

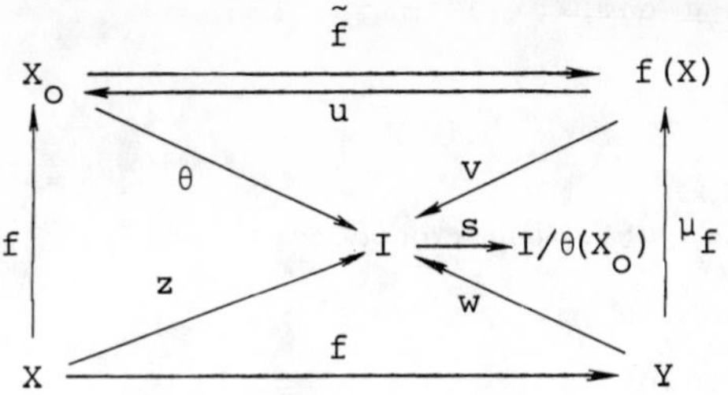

$X_o = X/f^{-1}(0)$,
$I = I(X_o)$, $\theta = \theta_{X_o}$
(cf. (4,18));
$s: I \longrightarrow I/\theta(X_o)$
is the canonical quotient mapping.

Set $z = \theta\lambda_f$. Hence by $(\mathcal{P}, I(Z))$, there is $w: Y \longrightarrow I$ such that $wf = z$, i.e. $w\mu_f \tilde{f}\lambda_f = \theta\lambda_f$, but λ_f being epi, this gives

(9) $$w\mu_f \tilde{f} = \theta.$$

Define $v = w\mu_f$. Then

$$sv\tilde{f}\lambda_f = swf = sz = s\theta\lambda_f = 0.$$

Since $\tilde{f}\lambda_f = 0$, this gives $sv = 0$. Hence there is $u: f(X) \longrightarrow X_o$ such that $\theta u = v = w\mu_f$. This combined with (9) yields

$$\theta u\tilde{f} = w\mu_f \tilde{f} = \theta,$$

hence $u\tilde{f} = 1_{X_o}$, for θ is mono. Since $\tilde{f}$ is always epi, the last equation proves that $X_o \cong f(X)$ (cf. [5], Chap. I, §1, Prop. 14).

The reader may wonder why we did not use a simple duality argument to prove (4,21). The reason for choosing the above proof is that it provides the simplest example of an "algebraic" way of looking at the duality problems in ℓ.c. spaces. Moreover, an easy generalization of this proof yields the above mentioned result of Palamodov:

(4,22) THEOREM [67]: *Let us consider a complex of morphisms*

$$W \xrightarrow{g} X \xrightarrow{f} Y$$

in TLC (i.e. $f\circ g = 0$). *Consider the following conditions*

(A) $\overline{\mathrm{Im}\, g} = \mathrm{Ker}\, f$ *and* f *is homo.*

(B) *Given* $S \neq \emptyset$, *the sequence of groups*

$$\text{(11)} \qquad \mathrm{Hom}_V(Y, \mathfrak{m}(S)) \xrightarrow{f^*} \mathrm{Hom}_V(X, \mathfrak{m}(S)) \xrightarrow{g^*} \mathrm{Hom}_V(W, \mathfrak{m}(S))$$

is exact. (Here we defined $f^*(v) = vf$, *etc.)*

Then (A) $\Longrightarrow$ (B) *for every* S. *Conversely, if* (B) *holds for at least one* S *such that* $\mathrm{card}(S) \geq \mathrm{card}((X/f^{-1}(0))')$, *then* (A) *holds.*

Roughly speaking this theorem says that the openness of a linear continuous mapping f can be "measured" by the exactness of sequences of groups (11). This clearly indicates that problems on TVS can be solved by algebraic methods. Needless to say, the foregoing theorem represents only a very simple example of links between homological algebra and functional analysis in *TLC*. Techniques of homological algebra become particularly efficient when applied to questions involving projective and inductive limits of *ℓ.c.* spaces. In particular, several results in Chapter 2 can be established and even strengthened by using homological methods (cf. [83, 84]). For further information on this subject the reader is referred to [66, 67, 82, 83, 84][1].

---

1) cf. also an interesting article of V.S.Retakh, *On the adjoint homomorphism in locally convex spaces*, Funkc. Analiz Prilož, 4 (1969), 63-71.

Stevens Institute of Technology
Castle Point Station, Hoboken
New Jersey 07030, U.S.A.

and

Universidade Federal de Pernambuco
Instituto de Matemática
Recife, Brasil

REFERENCES

[1] ARENS, R., EELLS, J., *On embedding uniform and topological spaces*, Pacif. J. Math., 6(1956), 397-403.

[2] BERENSTEIN, C.A., DOSTAL, M.A., "Analytically Uniform Spaces and their Applications to Convolution Equations", Lect. Notes in Math., vol.256, Springer-Verlag, 1972.

[3] BERENSTEIN, C.A., DOSTAL, M.A., *On convolution equations I*, "L'Analyse Harmonique dans le Domaine Complexe", Lect. Notes Math., vol.336, Springer-Verlag, 1973, 79-94.

[4] BOURBAKI, N., "Éléments de Mathématiques", Livre III, Topologie, Hermann, Paris, 1953-1961.

[5] BOURBAKI, N., "Éléments de Mathématiques", Livre V, Espaces Vectoriels Topologiques, Hermann, Paris, 1953 & 1955.

[6] BOURBAKI, N., "Éléments de Mathématiques", Livre VI, Intégration, Hermann, Paris, 1952-1963.

[7] BROWDER, F.E., *Functional analysis and partial differential equations I*, Math. Ann., 138(1959), 55-79.

[8] BUCUR, I., DELEANU, A., with the collaboration of P.J. Hilton, "Introduction to the theory of categories and functors", Wiley-Interscience, 1968.

[9] COLLINS, H.S., *Completeness and compactness in linear topological spaces*, Trans. Amer. Math. Soc., 79(1955), 256-280.

[10] COLLINS, H.S., *Completeness, full completeness and k-spaces*, Proc. Amer. Math. Soc., 6(1955), 832-835.

[11] COLLINS, H.S., *On the space* $\ell^\infty(S)$, *with the strict topology*, Math. Z., 106(1968), 361-373.

[12] DE WILDE, M., *Sur un type particulier de limite inductive*, Bull. Soc. Royale Sciences Liège, 35(1966), 545-551.

[13] DE WILDE, M., *Quelques théorèmes d'extension de fonctionnelles linéaires*, Bull. Soc. Royale Sciences Liège 35(1966), 552-557.

[14] DE WILDE, M., *Une propriété de relèvement des espaces à réseaux absorbants*, C.R. Acad. Sc. Paris, Série A, **266**(1968), 457-459.

[15] DE WILDE, M., *Réseaux dans les espaces linéaires à seminormes*, Mém. Soc. Royale Sciences Liège, **18**(1969), 7-144.

[16] DIEUDONNÉ, J., SCHWARTZ, L., *La dualité dans les espaces* (F) *et* (LF), Ann. Inst. Fourier, **1**(1950), 61-101.

[17] DOSTAL, M.A., *A complex characterization of the Schwartz space* $\mathcal{D}(\Omega)$, Math. Ann., **195**(1972), 175-191.

[18] DOSTAL, M.A., *On a class of complex* Λ*-structures*, to appear.

[19] DULST, D. van, *A note on* B- *and* B_r*-completeness*, Math. Ann., **197**(1972), 197-202.

[20] DULST, D. van, *On the non-equivalence of* B- *and* B_r*-completeness for strong duals of nuclear strict* (LF)*-spaces*, to appear.

[21] DULST, D. van, *Pathology in* $\mathcal{D}(\Omega)$ *and* $\mathcal{D}'(\Omega)$, to appear.

[22] EDWARDS, R.E., "Functional Analysis: Theory and Applications", Holt, Rinehart and Winston, New York, 1965.

[23] EHRENPREIS, L., *Solution of some problems of division III*, Amer. Math. J., **78**(1956), 685-715.

[24] EHRENPREIS, L., *Solution of some problems of division IV*, Amer. Math. J., **82**(1960), 522-588.

[25] EHRENPREIS, L., "Fourier Analysis in Several Complex Variables", Wiley-Interscience, London-New York, 1970.

[26] FIODOROVA, V.P., *A dual characterization of the complete hull and completeness of a uniform space*, Mat. Sbornik, **64**(1964), 631-639.

[27] FOIAŞ, C., MARINESCU, G., *Fonctionnelles linéaires dans les réunions dénombrables d'espaces de Banach réflexifs*, C.R. Acad. Sc. Paris, Série A, **261**(1965), 4958-4960.

[28] FREYD, P., "Abelian Categories", Harper & Row, New York - Evanston - London, 1964.

[29] GARNIR, H.G., DE WILDE, M., SCHMETS, J., "Analyse Fonctionnelle (Théorie constructive des espaces linéaires à semi-normes)", vol.I, Birkhäuser-Verlag, Basel und Stuttgart, 1968.

[30] GEĬLER, V.A., *On projective objects in the category of locally convex spaces*, Funkc. Analiz Prilož.,6(1972), 79-80.

[31] GROTHENDIECK, A., *Sur la complétion du dual d'un espace vectoriel localement convexe*, C.R. Acad. Sc. Paris, Série A, 230(1950), 605-606.

[32] GROTHENDIECK, A., *Sur les espaces* (F) *et* (DF), Summa Bras. Math., 3(1954), 57-123.

[33] GROTHENDIECK, A., "Produits tensoriels topologiques et espaces nucléaires", Memoirs Amer. Math. Soc., vol. 16, Providence, 1955.

[34] HARVEY, Ch., HARVEY, F.R., *Open mappings and the lack of fully completeness of* $\mathcal{D}'(\Omega)$, Bull. Amer. Math. Soc., 76(1970). 786-790.

[35] HÖRMANDER, L., *On the range of convolution operators*, Ann. of Math., 26(1962), 148-170.

[36] HORVÁTH, J., "Topological Vector Spaces and Distributions", Addison-Wesley, Reading, Mass., 1966.

[37] HUSAIN,T., "The Open Mapping and Closed Graph Theorems in Topological Vector Spaces", Clarendon Press, Oxford, 1965.

[38] JOHN, K., *Differentiable manifolds as topological linear spaces*, Math. Ann., 186(1970), 177-190.

[39] JOHN, K., DOSTAL, M., *Completion of certain* Λ*-structures*, Comment. Math. Univ. Carol., 7(1966), 93-103.

[40] KAKUTANI, S., *Concrete representation of abstract* (M)*-spaces*, Ann. of Math., 42(1941), 994-1024.

[41] KANTOROVITCH, L.V., AKILOV, G.P., "Functional Analysis in Normed Spaces", Pergamon Press, Oxford, 1964.

[42] KASCIC, M.J., Jr., ROTH, B., *A closed subspace of* $\mathcal{D}(\Omega)$ *which is not an* (LF)*-space*, Proc. Amer. Math. Soc., 24(1970), 801-803.

[43] KASCIC, M.J., Jr., *Functional analytic equivalences for* P *and strong* P*-convexity*, to appear.

[44] KATĚTOV, M., *On a category of spaces*, "Proc. First Symposium on General Topology, Prague 1961", Prague, 1962, 226-229.

[45] KATĚTOV, M., *On certain projectively generated continuity structures*, "Celebrazioni archimedee del secolo XX. Simposio di topologia", 1964, 47-50.

[46] KATĚTOV, M., *Projectively generated continuity structures: a correction*, Comment. Math. Carol., 6(1965), 251-255.

[47] KELLEY, J.L., *Hypercomplete linear topological spaces*, Mich. Math. J., 5(1958), 235-246.

[48] KELLEY, J.L., *Banach spaces with the extension property*, Trans. Amer. Math. Soc., 72(1952), 323-326.

[49] KELLEY, J.L., NAMIOKA, I., and collab., "Linear Topological Spaces", D, Van Nostrand, New York - London - Toronto, 1963.

[50] KÖTHE, G., *Die Quotientenräume eines linearen volkommenen Raumes*, Math. Z., 51(1947), 17-35.

[51] KÖTHE, G., *Hebbare lokalkonvexe Räume*, Math. Ann., **165**(1966), 181-195.

[52] KÖTHE, G., *Fortsetzung linearer Abbildungen lokalkonvexen Räume*, Jahresb. D.M.V., **68**(1966), 193-204.

[53] KÖTHE, G., "Topological Vector Spaces", Springer-Verlag New York Inc., 1969.

[54] KREĬN, M., ŠMULIAN, V., *On regularly convex sets in the space conjugate to a Banach space*, Ann. of Math., **41**(1940), 556-583.

[55] MAKAROV, B.M., *On inductive limits of normed spaces*, Vest. Leningr. Univ., 13(1965), 50-58.

[56] MALGRANGE, B., *Existence et approximation des solutions des équations aux dérivées partielles et des équations de convolution*, Ann. Inst. Fourier, 6(1955-1956), 271-355.

[57] MANES, E.G., *C(X) as a dual space*, Can. J. Math, **24**(1972), 485-491.

[58] MARKOV, A.A., *On free topological groups*, Dokl. AN SSSR, 31(1941), 299-302.

[59] MARKOV, A.A., *On free topological groups*, Izv. AN SSSR, ser. mat., **9**(1945), 3-64.

[60] MARTINEAU, A., *Sur le théorème du graphe fermé*, C.R. Acad. Sc. Paris, Série A, **263**(1966), 870-871.

[61] MARTINEAU, A., *Sur des théorèmes de S.Banach et L.Schwartz concernant le graphe fermé*, Studia Math., 30(1968), 43-51.

[62] MICHAEL, E., *Three mapping theorems*, Proc. Amer. Math. Soc., 15(1964), 410-415.

[63] MICHAEL, E., *A short proof of the Arens-Eells embedding theorem*, Proc. Amer. Math. Soc., 15(1964), 415-416.

[64] NACHBIN, L., *A theorem of Hahn-Banach type for linear transformations*, Trans. Amer. Math. Soc., 68(1950), 28-46.

[65] NACHBIN, L., *Some problems in extending and lifting continuous linear transformations*, "Proc. Intern. Symposium Lin. Spaces, Jerusalem, 1960", Jerusalem, 1961, 340-350.

[66] PALAMODOV, V.P., *The functor of projective limit in the category of topological linear spaces*, Mat. Sbornik, 75(1968), 567-603.

[67] PALAMODOV, V.P., *Homological methods in the theory of locally convex spaces*, Uspekhi Mat. Nauk, 26(1971), 3-65.

[68] PIETSCH, A., "Nukleare lokalkonvexe Räume", Akademie-Verlag, Berlin, 1965.

[69] PTÁK, V., *On complete topological linear spaces*, Čechosl. Mat. Ž., 78(1953), 301-364.

[70] PTÁK, V., *Weak compactness in convex topological vector spaces*, Czechosl. Math. J., 79(1954), 175-186.

[71] PTÁK, V., *Completeness and the open mapping theorem*, Bull. Soc. Math. France, 86(1958), 41-74.

[72] PTÁK, V., *A combinatorial lemma on the existence of convex means and its application to weak compactness*, Proc. Symp. Pure Math., Amer. Math. Soc., 7(1963), 437-450.

[73] PTÁK, V., *Some open mapping theorems in* LF-*spaces and their application to existence theorems for convolution equations*, Math. Scand., 16(1965), 75-93.

[74] PTÁK, V., *Algebraic extensions of topological spaces*, "Contributions to extension theory of topological structures", (Proceedings of the Symposium, Berlin 1967), Deutsche Verlag Wiss., 1969, 179-188.

[75] PTÁK, V., *Openness of linear mappings in* LF-*spaces*, Czech. Math. J., 94(1969), 547-552.

[76] PTÁK, V., *Simultaneous extensions of two functionals*, Czech. Math. J., **19**(1969), 553-566.

[77] PTÁK, V., *Extension of sequentially continuous functionals of inductive limits of Banach spaces*, Czech. Math. J., **95**(1970), 112-121.

[78] RAĬKOV, D.A., *Free locally convex spaces of uniform spaces*, Math. Sbornik, **63**(1964), 582-590.

[79] RAĬKOV, D.A., *Double closed-graph theorem for topological linear spaces*, Siber. Math. J., 7(1966), 287-300.

[80] RAĬKOV, D.A., *Closed-graph and open-mapping theorems*, Appendix 1 to the Russian edition of [86], Moscow, 1967.

[81] RAĬKOV, D.A., *Some linear topological properties of spaces* $\mathcal{D}$ *and* $\mathcal{D}'$, Appendix 2 to the Russian edition of [86], Moscow, 1967.

[82] RAĬKOV, D.A., *Semiabelian categories*, Dokl. AN SSSR, **188**(1969), 1006-1009.

[83] RETAKH, V.S., *On the dual of a subspace of a countable inductive limit*, Dokl. AN SSSR, **184**(1969), 44-46.

[84] RETAKH, V.S., *On subspaces of a countable inductive limit*, Dokl. AN SSSR, **194**(1970), 1277-1279.

[85] ROBERTSON, A.P., ROBERTSON, W., *On the closed graph theorem*, Proc. Glasgow Math. Assoc., 3(1956), 9-12.

[86] ROBERTSON, A.P., ROBERTSON, W., "Topological Vector Spaces", Cambridge Univ. Press, Cambridge - New York, 1964.

[87] ROBERTSON, W., *Completion of topological vector spaces*, Proc. Lond. Math. Soc., 8(1958), 242-257.

[88] ROBERTSON, W., *On the closed graph theorem and spaces with webs*, Proc. Lond. Math. Soc., **24**(1972), 692-738.

[89] SCHAEFFER,H.H., "Topological Vector Spaces", Mac Millan, New York, 1966.

[90] SCHWARTZ, L., *Sur le théorème du graphe fermé*, C.R. Acad. Sc. Paris, Série A, 263(1966), 602-605.

[91] SŁOWIKOWSKI, W., *On continuity of inverse operators*, Bull. Amer. Math. Soc., **67**(1961), 467-470.

[92] SŁOWIKOWSKI, W., *Quotient spaces and the open mapping theorem*, Bull. Amer. Math. Soc., **67**(1961), 498-500.

[93] SŁOWIKOWSKI, W., *Fonctionnelles linéaires dans des réunions dénombrables d'espaces de Banach réflexifs*, C.R. Acad. Sc. Paris, Série A, 262(1966),870-872.

[94] SŁOWIKOWSKI, W., "Epimorphisms of adjoints to generalized (LF)-spaces", Aarhus Universitet, Matematisk Institut, Lecture Notes, 1966.

[95] SŁOWIKOWSKI, W., *Range of operators and regularity of solutions*, Studia Math., **40**(1971), 183-198.

[96] SŁOWIKOWSKI, W., *On Hörmander's theorem about surjections of* $\mathcal{D}'$, to appear in Math. Scand.

[97] SMOLJANOV, O.G., *Almost closed linear subspaces of strict inductive limits of sequences of Frēchet spaces*, Mat. Sbornik, **80**(1969), 513-520.

[98] SUMMERS, W.H., *Products of fully complete spaces*, Bull. Amer. Math. Soc., 75(1969), 1005.

[99] SUNYACH, C., *Sur le thēorēme du graphe fermēe*, "Séminaire P. Lelong (Analyse), 8e Année, 1967-1968", Lect. Notes in Math., vol. 71, Springer-Verlag, 1968, 33-37.

[100] TOMÁŠEK, S., *On a theorem of M.Katětov*, Comment. Math., Univ. Carol., 7(1966), 105-108.

[101] TOMÁSEK, S., *Über eine lokalkonvexe Erweiterung von topologischen Produkten*, "Contributions to extension theory of topological structures", (Proceedings of the Symposium, Berlin, 1967), Deutsche Verlag. Wiss., 1969, 233-237.

[102] TOMÁŠEK, S., *Certain generalizations of the Katetov theorem*, Comment. Math. Univ. Carol., **9**(1968), 573-581.

[103] TOMÁŠEK, S., *On a certain class of* Λ*-structures I*, Czechosl. Math. J., 20(1970), 1-17.

[104] TOMÁŠEK, S., *On a certain class of* Λ*-structures II*, Czechosl. Math. J., 20(1970), 19-33.

[105] TRÈVES, F., "Linear Partial Differential Equations with Constant Coeficients", Gordon & Breach, New York, 1966.

[106] TRÈVES, F., "Topological Vector Spaces, Distributions and Kernels", Acad. Press, New York and London, 1967.

[107] TRÈVES, F., "Locally Convex Spaces and Linear Partial Differential Equations", Springer-Verlag New York Inc., 1967.

Added in the proofs:

[108] SHAVGULIDZE; Je. T., *On the hypercompleteness of locally convex spaces*, Mat. Zametki, **13**(1973), 297-302.

[109] SMOLJANOV, O.G., *The space* $\mathfrak{D}$ *is not hereditarily complete*, Math. USSR Izvestija, 5(1971), 696-710.

[110] SUMMERS, W.H., *Full-completeness in weighted spaces*, Can. J. Math., **22**(1970), 1196-1207.

USEFULNESS OF PSEUDODIFFERENTIAL AND FOURIER INTEGRAL OPERATORS IN THE STUDY OF THE LOCAL SOLVABILITY OF LINEAR PARTIAL DIFFERENTIAL EQUATIONS

by
F. Trèves

I will try to show how the theory of pseudodifferential and Fourier integral operators has been of help in studying the problem of the local solvability of a linear partial differential equation *with simple real characteristics*,

$$P(x,D)u = f. \tag{1}$$

At the end of this lecture I shall give a statement, in a form somewhat different from the usual one, of the so-called Condition ($\mathcal{P}$) which has been recently proved to be sufficient, and to a large extent necessary, in order that Eq. (1) be locally solvable at a given point x_o.

We use the standard notation

$$D = (D_{x^1},\dots,D_{x^N}), \quad D_{x^j} = -\sqrt{-1}\,\frac{\partial}{\partial x^j}.$$

We shall also write $\partial_{x^j} = \partial/\partial x^j$. We assume throughout, that $P(x,D)$ is a linear partial differential operator of order $m > 0$, with C^∞ coefficients, in an open subset Ω of the Euclidean space $\mathbb{R}^N$. One could, and should, for a number of reasons, also consider the case of a pseudodifferential equation. But we shall limit ourselves mainly to differential equations. Now, one says that the equation (1) is *locally solvable at* $x_o \in \Omega$ if there is an open neighborhood U of x_o in Ω such that, given any C^∞ function f in $\overline{U}$, the closure of U, there is a distribution $u \in \mathcal{D}'(U)$ such that (1) holds in U. That $f \in C^\infty(\overline{U})$ means that f can be extended as a C^∞ function in a neighborhood of $\overline{U}$ (we are assuming, which we obviously may, that the boundary of U is sufficiently regular).

We have said that the differential operator $P(x,D)$ should have *simple real characteristics* or, as it is often said, be of *principal type*. This is defined in terms of its principal part $P_m(x,D)$, which is homogeneous of degree m with respect to D and such that

$$P(x,D) - P_m(x,D)$$

is a differential operator of order $<m$. We consider then the *principal symbol* $P_m(x,\xi)$ of $P(x,D)$: it is a homogeneous polynomial of degree m with respect to $\xi = (\xi_1,\ldots,\xi_N)$ with coefficients which are complex-valued C^∞ functions in Ω. One says that $P(x,D)$ is of *principal type* if $d_\xi P_m(x,\xi)$ does not vanish for any x $\Omega \in$ and any $\xi \in \mathbb{R}_N$, $\xi \neq 0$. The geometric meaning of this condition is the following: by Euler's homogeneity relation we have

$$P_m(x,\xi) = \frac{1}{m}\, \xi \cdot d_\xi P_m(x,\xi)$$

(the dot denotes the usual scalar product). Therefore $d_\xi P_m(x,\xi) = 0$ implies $P_m(x,\xi) = 0$, and thus if $P(x,D)$ is of principal type, wherever $P_m(x,\xi)$ vanishes, provided that we exclude the points of the form $(x,0)$, at least of its ξ-derivatives should not vanish. Consider then the set of points $\xi \neq 0$, $C_P(x)$, such that $P_m(x,\xi) = 0$. It is a *cone* in $\mathbb{R}_N$. By definition, $P(x,D)$ is of principal type if, given any x in Ω, the cone $C_P(x)$ has only simple points (or simple generators). This motivates the terminology "simple real characteristics" since the solutions (x,ξ) of $P_m(x,\xi)$ are called characteristic points, $C_P(x)$ is called the characteristic cone at the point x in the base Ω, the union C_P of the $C_P(x)$ as x ranges over is called the characteristic set of $P(x,D)$. It should be pointed out that the correct setting for these notions is the *cotangent bundle* $T^*(\Omega)$ over Ω. Indeed, $P_m(x,\xi)$ is a C^∞ function in $T^*(\Omega)$ and C_P is a conic subset of $T^*(\Omega)$ (conic means that the intersection of C_P with every cotangent space $T^*_x(\Omega)$ at a point x is a cone).

To start our "attack" on the problem of the solvability of Eq. (1) we begin by localizing in the cotangent bundle $T^*(\Omega)$ (the reader unfamiliar with vector bundles may identify $T^*(\Omega)$ with the product $\Omega \times \mathbb{R}_N$). That we are able to take advantage of such localizations is entirely due to the theory of *pseudodifferential operators*, and would have not been possible in the "classical" framework.

Let (x_o,ξ^o) be a point of the complement $\dot{T}^*(\Omega)$ of the zero section in $T^*(\Omega)$ ($\dot{T}^*(\Omega)$ can be identified with $\Omega \times (\mathbb{R}_N \setminus \{0\})$). Assume first that

(2) $$P_m(x_o,\xi^o) \neq 0.$$

We can find a C^∞ function $g(x,\xi)$ in $\dot{T}^*(\Omega)$, positive-homogeneous of degree *zero* (i.e., $g(x,\rho\xi) = g(x,\xi)$ for all x, ξ and $\rho > 0$), equal to one in some neighborhood of (x_o,ξ^o), moreover with support entirely contained in the region where $P_m(x,\xi) \neq 0$. This means that, for a suitable constant $c_o > 0$, we have, on the support of $g(x,\xi)$ (chose projection in the base Ω we shall assume to be compact),

(3) $$|P_m(x,\xi)| \geq c_o |\xi|^m.$$

In the standard terminology, this means that $P(x,D)$ is *elliptic* in a neighborhood of the support of $g(x,\xi)$. Pseudodifferential operators provide us with a mechanical procedure for solving (approximately) the "localized" equation:

(4) $$P(x,D)v = g(x,D)f,$$

where

(5) $$g(x,D)f(x) = (2\pi)^{-N} \int e^{ix\cdot\xi} g(x,\xi)\hat{f}(\xi)d\xi.$$

We have used here the standard notation: $\hat{f}$ is the Fourier transform of f, which we assume to have compact support (possibly after some extension beyond a given neighborhood of x_o); $x\cdot\xi = x^1\xi_1 + \dots + x^N\xi_N$. For the technique of solving (4), I can only refer to the texts on the theory of pseudodifferential operators, or to my own lectures in this same conference on that subject.

Suppose now that we have:

(6) $$P_m(x_o,\xi^o) = 0.$$

This is where the principal type hypothesis will be of help to us. We know that some ξ-derivative of $P_m(x,\xi)$ does not vanish at (x_o,ξ^o). Let us assume that it is the partial derivative with respect to ξ_N. By the implicit functions theorem we can then write

(7) $$P_m(x,\xi) = Q(x,\xi)(\xi_N - \lambda(x,\xi'))$$

is a *conic* open neighborhood U of (x_o,ξ^o) in $\dot{T}^*(\Omega)$. Both $Q(x,\xi)$ and $\lambda(x,\xi')$ are C^∞ functions in U; we may take $Q(x,\xi)$ to be homogeneous in ξ of degree $m-1$, and λ homogeneous in ξ' of degree 1. Furthermore, if we choose U suitable "narrow" around its axis (which is the "half-ray" through (x_o,ξ^o)), we may assume that, for a sufficiently small constant $c_1 > 0$,

(8) $$|Q(x,\xi)| \geq c_1|\xi|^{m-1}, \quad \forall (x,\xi) \in U.$$

Let $g(x,\xi)$ as before, except that we require now its support to lie in the conic open set U, and not in the region where (3) is valid. It is not difficult to check that, up to a pseudodifferential operator of order $-\infty$, we have:

(9) $$g(x,D)P(x,D) = g(x,D)Q(x,D)L(x,D),$$

where

(10) $$L(x,D) = D_N - \lambda(x,D') - c(x,D),$$

with $c(x,D)$ a pseudodifferential operator (in Ω) of order at most zero. By the technique of inversion of elliptic pseudodifferential operators, we may solve approximately the equation

(11) $$Q(x,D) = g_1(x,D)f,$$

an then try to solve (again approximately) the equation

$$(12) \qquad L(x,D)v = g_2(x,D)w.$$

The functions g_1 and g_2 are exactly like g, except that $g_1 \equiv 1$ in a neighborhood in $\dot{T}^*(\Omega)$ of the support of g, whereas $g_2 \equiv 1$ in a neighborhood of the support of g_1 (all the supports must be contained in $\mathcal{U}$). By combining the equations (11) and (12) we obtain an approximate solution of Eq. (4).

If U is a sufficiently small neighborhood of x_o in Ω, we may find an open covering of $\dot{T}^*(U)$ consisting of a *finite* number of conic sets $\mathcal{U}$ in which either we have (3) or we have a factorization of the kind (7), with all the properties we have indicated, in particular (8). We can take the cut-off function $g(x,\xi)$ ranging over a partition of unity subordinate to this covering and define u as the sum of the solutions v of the corresponding equations (4). It is easy to check that u is a solution of the equation:

$$(13) \qquad P(x,D)u = f + Rf,$$

where R is a pseudodifferential operator of order -1 (we take now $f \in C_c^\infty(U)$, which can of course always be achieved by a redefinition of U and f). Moreover, if we have done all that is needed, we may express u in the form $u = Kf$, where K is a bounded linear operator on $L^2(U)$ (as the anlysis shows, we can even obtain that K be a bounded linear operator of $L^2(U)$ into $H^{m-1}(U)$ with norm decreasing to zero with the diameter of U). It is well known that the norm of R, viewed as a bounded linear operator on $L^2(U)$, also decreases to zero with the diameter of U. Thus, if the latter is sufficiently small, the operator $I+R$ is invertible, and we obtain an exact solution of (1) by writing:

$$P(x,D)\{K(I+R)^{-1}f\} = f. \tag{14}$$

Summarizing, the theory of pseudodifferential operators has enabled us to reduce the solvability problem of Eq. (1) to that of the first-order pseudodifferential equations (12). A further reduction, of crucial significance, will now be made possible by the theory of *Fourier integral operators*.

It is convenient, and instructive to view equations such as (12) as *evolution equations*. Let us change variables, and set $t = x^N - x_o^N$, $y^j = x^j$ for $j \leq N-1 = n$ (the covariables ξ_j, $j < N$, will be denoted by η_j and ξ_N by τ). In this notation we are dealing with equations of the general type

$$D_t u - \lambda(y,t,D_y)u - c(y,t,D_y,D_t)u = f, \tag{15}$$

where $\lambda(y,t,\eta)$ is homogeneous of degree 1 in η and $c(y,t,\eta,\tau)$ is a symbol of degree < 0. Both these functions can be assumed to be C^∞ with respect to all arguments in $\mathbb{R}^{n+1} \times (\mathbb{R}_{n+1} \setminus \{0\})$ (this tacitely assumes that we have extended them to the whole space and multiplied them by suitable cut-off functions; this does not affect the preceding argument, it merely introduces here and there a few more error terms, expressed by pseudodifferential operators of order $-\infty$). We may even assume that the (y,t)-projections of the supports of $\lambda(y,t,\eta)$ and $c(y,t,\eta,\tau)$ are compact. Then we know that the pseudodifferential opertor $c(y,t,D_y,D_t)$ defines a bounded linear operator of each Sobolev espace $H^s(\mathbb{R}^{n+1})$ into itself, and because of this can easily eliminated from the picture. We may concentrate our attention upon the equation

$$Lu = D_t u - \lambda(y,t,D_y)u = f. \tag{16}$$

We shall write $\lambda = a + \sqrt{-1}\, b$, where a and b are *real-valued* C^∞ functions of (y,t,η), $\eta \neq 0$, homogeneous of degree one in η. Observe that the pseudodifferential operator $a(y,t,D_y)$ is an operator on distributions in the y-variables, depending smoothly on t. When so viewed, let us denote it by $A(t)$; similarly, denote $b(y,t,D_y)$ by $B(t)$. If we replace f by $-\sqrt{-1}\, f$, the equation (16) reads:

$$\partial_t u - \sqrt{-1}\, A(t)u + B(t)u = f. \tag{17}$$

The theory of Fourier integral operators will now enable us to get rid of the therm $\sqrt{-1}\, A(t)$. Observe that since both a and b are *real*, the operators $A(t)$ and $B(t)$, defined on the dense subset $H^1(\mathbb{R}^n_y)$ of $H^0(\mathbb{R}^n_y)$, are equal to two self-adjoint (unbounded) linear operators on $H^0(\mathbb{R}^n_y)$, at least modulo bounded linear operators. Suppose for a moment that A and B are independent of t. The preceding observation allows us to solve the operator equation:

(18) $$\partial_t U = iAU,$$

with initial condition:

(19) $$U\Big|_{t=0} = I, \text{ the identity operator.}$$

As a matter of fact, the solution is well konwn: $U(t) = \exp(iAt)$, the group of unitary operators on $L^2(\mathbb{R}^n)$ with infinitesimal generator A (when A is self-adjoint; when A is merely self-adjoint modulo bounded linear operators, $U(t)$ is "almost unitary".)

Set then $u = U(t)v$, $f = U(t)g$ in Eq. (17), where A and B are independent of t. It gets transformed into

(20) $$\partial_t v + U(t)^{-1}BU(t)v = g.$$

It is checked at once that, up to a bounded linear operator (depending smoothly on t),

(21) $$B^{\#}(t) = U(t)^{-1}BU(t)$$

is self-adjoint. If we set, as usually done in Lie groups and Lie algebras theory,

(22) $$(\mathrm{Ad}\ A)(B) = [A,B] = AB - BA,$$

we see easily that

$$B^{\#}(t) = e^{-iAt}Be^{iAt} = e^{-it(\mathrm{Ad}\ A)}B \tag{23}$$

$$= \sum_{j=0}^{+\infty} \frac{1}{j!}(-it)^j(\mathrm{Ad}\ A)^j B.$$

These formulae have a geometrical "substratum", as we well know. Suppose for instance that both A and B are first-order differential operators with real coefficients (i.e., real vector fields) multiplied by $-\sqrt{-1}$:

$$A = \sum_{k=1}^{n} a^k(y)D_{y^k}, \qquad B = \sum_{k=1}^{n} b^k(y)D_{y^k}.$$

Consider then the solution $z = z(y,t)$ of the system of ordinary differential equations:

$$\frac{dz^k}{dt} = -a^k(z), \quad 1 \leqslant k \leqslant n, \tag{24}$$

with initial conditions:

$$z^k\Big|_{t=0} = y^k, \quad 1 \leqslant k \leqslant n. \tag{25}$$

For small values of $|t|$,

$$z = z(y,t) \tag{26}$$

defines a C^∞ change of variables in the neighborhood of any given point. Let

(27) $$y = \mathcal{Y}(z,t)$$

denote the inverse change of variables. We note that $t \longrightarrow (\mathcal{Y}(z,t),t)$ is the (piece of) integral curve of the vector field $\partial_t - iA$ through the point z. The change of variables $(y,t) \longmapsto (z,t)$ transforms that vector field into ∂_t. An easy computation shows that

(28) $$B^\#(t) = \sum_{k,\ell=1}^{n} b^k(\mathcal{Y}(z,t)) \frac{dz^\ell}{dy^k} D_{z^\ell}.$$

Observe that $b^k(\mathcal{Y}(z,t))$ is the value of b^k at the point reached at time t when we move along the curve $t \longmapsto \mathcal{Y}(z,t)$ (z is the point of this curve obtained at $t = 0$). Note also that the symbol of $B^\#(t)$ is

(29) $$b(\mathcal{Y}(z,t), {}^t\!\left(\frac{\partial z}{\partial y}\right)\zeta),$$

where $\frac{\partial z}{\partial y}$ stands for the Jacobian matrix of the z's with respect to the y's. The symbol (29) is nothing else but the transform of the symbol $b(y,\eta)$ of B under the mapping $(z,\zeta) \longmapsto (y,\eta)$, where

(30) $$y = \mathcal{Y}(z,t), \quad \eta = {}^t\!\left(\frac{\partial z}{\partial y}\right)(y,t).$$

Now, (30) is the transformation in the cotangent bundle (over $\mathbb{R}^n$ or over an open subset of $\mathbb{R}^n$) associated with the transformation (27) in the base. The curve $t \longmapsto \mathcal{Y}(z,t)$ in the base is the projection of the curve $t \longmapsto (\mathcal{Y}(z,t), {}^t(\partial z/\partial y)\zeta)$ in the cotangent bundle, and the symbol of $B^{\#}(t)$ is obtained by displacing that of B along the latter curve. This describes completely the transformation of B into $B^{\#}(t)$, and shows that introducing the cotangent bundle was not just a dressing-up in fancy language of otherwise plain material, but had to do with really deep aspects of the problem.

In the more general situation, when A and B are not vector fields on an open set of $\mathbb{R}^n$, they can still be regarded as vector fields on an appropriate Lie group (they can be regarded as elements of its Lie algebra) and $B^{\#}(t)$ can once more be interpreted as a "displacement" of B. However, when, as it is the case in our problem, they are defined by pseudodifferential operators of order one, a more concrete interpretation of the whole operation is possible, thanks to Fourier integral operators.

Let us therefore go back to the case where

$$A(t) = a(y,t,D_y), \quad B(t) = b(y,t,D_y)$$

are essentially self-adjoint paseudodifferential operators of order one. Formally the process is the same as that which lead us from Eq. (17) to Eq. (2)): we solve the Cauchy problem (18)-(19) and introduce the transform $B^{\#}(t)$ (note only that $B = B(t)$ now). It is not difficult to prove that $U(t)$ is essentially unitary (on $L^2(\mathbb{R}^n)$), and that $B^{\#}(t)$, like $B(t)$, is essentially self-adjoint. But the question for us is whether $B^{\#}(t)$ is also a pseudodifferential operator of order one, to which some appropriate analysis can be applied. The beauty of the approach is that this is indeed so and that, moreover, $B^{\#}(t)$ can be related to $B(t)$ in a simple and elegant manner, generalizing what happens when A and B are vector fields. This is due to the fact that $U(t)$ can be approximated, modulo regularizing operators, by a Fourier integral operator:

$$(31) \qquad K(t)u(y) = (2\pi)^{-n} \int e^{i\phi(y,t,\eta)} k(y,t,\eta)\hat{u}(\eta)d\eta ,$$

where $k(y,t,\eta)$, the *amplitude-function*, is a symbol of degree zero (depending smoothly on t) and the *phase-function* ϕ is C^{∞} with respect to (y,t,η), $\eta \neq 0$, homogeneous of degree one with respect to η, and *real-valued*. In fact, it is the unique solution of the following (nonlinear) first-order Cauchy problem:

$$(32) \qquad \phi_t = a(y,t,\phi_y), \qquad \phi\Big|_{t=0} = y\cdot\eta .$$

Assuming that we have added a zero-order term to $a(y,t,D_y)$ so as to make it self-adjoint, we can obtain that $U(t)^{-1}$ be equal to the adjoint $K(t)^*$ of $K(t)$ modulo regularizing operators, and thus (again modulo regularizing operators)

$$(33) \qquad B^{\#}(t) \equiv K(t)^*B(t)K(t).$$

At this stage the important *Egorov's theorem* comes to our rescue: it states that K^*BK is a pseudodifferential operator of same order as B, here *one*, and that modulo symbols of order strictly less, its symbol can be computed out of that of B by a formula similar, and generalizing the one which yielded (29). The relevant curves in the cotangent bundle are now the *bicharacteristic strips* of $\tau - a(y,t,\eta)$, i.e., of the symbol of $\frac{1}{i}(\partial_t - iA)$. These strips are the integral curves of the *Hamiltonian* vector field of $\tau - a(y,t,\eta)$, which is the vector field

$$\mathcal{H} = \partial_t - a_\eta(y,t,\eta)\cdot\partial_y + a_y(y,t,\eta)\cdot\partial_\eta \,.$$

They are the curves described by the point $(z(y,t,\eta),t,\zeta(y,t,\eta))$, where

$$\frac{dz}{dt} = -a_\eta(z,t,\zeta), \quad \frac{d\zeta}{dt} = a_y(z,t,\zeta), \tag{34}$$

$$z = y, \quad \zeta = \eta \quad \text{at} \quad t = 0. \tag{35}$$

For small values of $|t|$, the mapping $(y,\eta) \longmapsto (z(y,t,\eta),\zeta(y,t,\eta))$ is a diffeomorphism; let us denote by

$$y = \mathcal{Y}(z,t,\zeta), \quad \eta = \eta(z,t,\zeta), \tag{36}$$

the inverse transformation. If $b^\#(y,t,\eta)$ denotes the symbol of $B^\#(t)$, we have:

(37) $b^{\#}(z,t,\zeta) \equiv b(\mathcal{Y}(z,t,\zeta),t,\eta(z,t,\zeta))$ *modulo symbols of degree zero.*

The reader will check without too much difficulty that this generalizes the formula (28).

In such a manner have we reduced our original problem to the solvability of the evolution equation:

$$\partial_t u + b^{\#}(y,t,D_y)u = f, \tag{38}$$

where we have the right to assume that the *principal symbol* $b_o^{\#}(y,t,\eta)$ of $b^{\#}(y,t,D_y)$ is *real* (and is defined by Property (37)).

With the pseudodifferential equation (38) we associate the following ordinary differential equation, depending on the parameters (y,η):

$$\partial_t w + b^{\#}(y,t,\eta)w = \hat{f}(y,t,\eta). \tag{39}$$

Eq. (39) is first-order and linear. All solutions are known. They can be written in the form:

$$w(y,t,\eta) = \int_{T_o}^{t} e^{B(y,t,t',\eta)}\,\hat{f}(y,t',\eta)\,dt', \tag{40}$$

where

$$B(y,t,t',\eta) = \int_{t}^{t'} b^{\#}(y,t'',\eta)dt''. \tag{41}$$

We shall consider the right-hand sides $\hat{f}$ having compact support with respect to (y,t), contained in a small slab $|t| < T$, and tempered with respect to η. Then we ask whether we can always (i.e., given any such right-hand side $\hat{f}$) find a solution w of (39) which is *tempered* with respect to η at infinity. A simple argument shows that this property is equivalent to the following one:

(ψ) *There is a number* T_0, $|T_0| \leq T$, *such that, for every* $t, t' \in]-T,T[$ *such that* t' *lies in the segment joining* t *to* T_0, *for all* y, η,

$$B(y,t,t',\eta) \leq 0 \qquad \textit{(mod symbols of degree zero)}. \tag{42}$$

The existence of the number T_0 is, in turn, equivalent to the following condition:

($\tilde{\psi}$) *whatever* y, η, *if* $b^{\#}_0(y,t,\eta) < 0$ *for some* t, $|t| < T$, *then* $b^{\#}_0(y,t',\eta) \leq 0$ *for every* t', $t < t' < T$.

We may translate ($\tilde{\psi}$) in terms of $b(y,t,\eta)$ (which, we recall, is homogeneous of degree one with respect to η):

(Ψ) *along every bicharacteristic strip of* $\tau - a(y,t,\eta)$ (in a neighborhood of a point $(y_o,0,\eta^o,\tau^o)$), *if* $b(y,t,\eta)$ *is* < 0 *at some point, it remains* ≤ 0 *at every later point* (bicharacteristic strips are oriented curves).

But suppose that $b(y,t,\eta)$, restricted to a bicharacteristic strip Γ of $\tau - a(y,t,\eta)$, changes sign at some point (y_1,t_1,η^1,τ^1) of Γ. If

(Ψ) holds, it must necessarily change sign from + to -. If we now make the "symmetry" $(y,t,\eta,\tau) \longleftrightarrow (y,t,-\eta,-\tau)$ and look at the behavior of b along the bicharacteristic strip of $\tau - a(y,t,\eta)$ through $(y_{\cdot},t_1,-\eta^1,-\tau^1)$, we see that b changes sign there from - to +, and therefore violates (Ψ)! Thus, due to its homogeneity of degree one with respect to η, we see that (Ψ) is equivalent with:

($\mathcal{P}$) *along every bicharacteristic strip of* $\tau - a(y,t,\eta)$, *in a neighborhood of* $(y_o,0,\eta^o,\tau^o)$, $b(y,t,\eta)$ *does not change sign.*

This is in essence the solvability condition ($\mathcal{P}$) which, as R. Beals and C. Fefferman have shown, implies the local solvability of the original equation (1) (of course, for this it must be satisfied everywhere in the cotangent bundle over a neighborhood of the point $x_0 = (y_0,0)$ under consideration). We have seen that it means that the ordinary differential equation (39) has a tempered solution (tempered with respect to η) whenever the right-hand side is tempered.

Rutgers University
Department of Mathematics
New Brunswick, N.J. 08903
U.S.A.

BIBLIOGRAPHY

[1] R.BEALS and C.FEFFERMAN, *On local solvability of linear partial differential equations*, Ann. of Math., 97(1973),482-498.

[2] L.NIRENBERG and F.TRÈVES, *On local solvability of linear partial differential equations, Part I: Necessary Conditions.* Comm. Pure Applied Math., vol. XXIII, pp.1-38 (1970).

[3] F.TRÈVES, *On the existence and regularity of solutions of linear partial differential equations*, A.M.S. Summer School on partial differential equations, Berkeley (Calif.) 1971.

BOUNDING SETS IN BANACH SPACES AND REGULAR CLASSES OF ANALYTIC FUNCTIONS

by

Martin Schottenloher

INTRODUCTION

Closely related with continuation properties of analytic functions in the finite dimensional case is the notion of convexity with respect to a family of analytic functions. So, for a domain X in $\mathbf{C}^n$ the following three properties are equivalent:

X is holomorphically convex,

(*) X is a domain of holomorphy,

X is a domain of existence [8, p.43 ff and p.283].

Although the last two notions extend nearly without difficulties to the infinite dimensional case, it is not obvious under which conditions a given domain should be called holomorphically convex. The consideration of this question leads to the notion of the bounding set which was first introduced by ALEXANDER [1]: A subset B of a Banach space E is called a bounding set if every analytic function on E is bounded on B.

In the first section of this paper we show that for a large class of Banach spaces, including the separable and reflexive spaces, the bounding sets are the relatively compact sets. This is already known [6], [9] but we are interested in the different proof which allows to extend some classical results to the infinite dimensional case. This is done in the second section by studying regular classes of analytic functions and proving some characterizations similar to (*) with respect to regular classes. The paper ends with a survey of results on analytic continuation which can be obtained by applying the theory of regular classes.

1. BOUNDING SETS

A domain X in $\mathbf{C}^n$ is called *holomorphically convex* if one of the following three equivalent conditions is satisfied [8]:

1° - For every sequence (x_n) in X without accumulation point in X there exists $f \in \mathcal{H}(X)$ such that $\sup\{|f(x_n)| \mid n \in \mathbb{N}\} = \infty$.

2° - The holomorphically convex hull $\hat{K}$ of every compact set $K \subset X$ is compact.

3° - $d_X(\hat{K}) > 0$ for every compact set $K \subset X$.

Here $\mathcal{H}(X)$ denotes the algebra of analytic functions on X, $\hat{K}$ is defined by

$$\hat{K} = \{x \in X \mid |f(x)| \leqslant \|f\|_K \text{ for all } f \in \mathcal{H}(X)\},$$

where

$$\|f\|_K = \sup\{|f(y)| \mid y \in K\},$$

and $d_X(\hat{K})$ is the boundary distance of $\hat{K}$:

$$d_X(\hat{K}) = \inf\{\|y-z\| \mid y \in \hat{K}, \quad z \in \mathbf{C}^n \setminus X\}.$$

Now let X be a domain in the infinite dimensional complex Banach space E (general reference for the theory of analytic mappings is [12]). Because of $\hat{K} \subset C \circ K$ ($\equiv$ convex hull of K) the following implications hold: $1° \Longrightarrow 2° \Longleftrightarrow 3°$.

In the particular case $X = E$ the property 2º is always satisfied and 1º is equivalent to the condition: Every subset $B \subset E$ which satisfies $\|f\|_B < \infty$ for all $f \in \mathcal{H}(E)$ is relatively compact.

DEFINITION: A subset B of E will be called a *bounding set* if $\|f\|_B < \infty$ for all $f \in \mathcal{H}(E)$.

One is interested to classify those Banach spaces in which the bounding sets have compact closure.

PROPOSITION: *In a separable Banach space* E *every bounding set is relatively compact.*

PROOF [13]: Let $B \subset E$ be not relatively compact. Then there exist $r > 0$ and a sequence (x_n) in $\overline{B}$ such that $\|x_n - x_m\| \geq 2r$ for $n \neq m$. Let (a_j) be dense in E and define

$$K_n = \overline{C \circ \bigcup_{j \leq n} B(a_j, r)}$$

where $B(z,r)$ is the ball in E with radius r and center $z \in E$. Then i) $\overline{C \circ K_n} = K_n$ for all $n \in \mathbb{N}$ and ii) $E = \bigcup\{\overset{\circ}{K}_n \mid n \in \mathbb{N}\}$. K_n can be covered by a finite collection of balls with radius 2r. Hence it can be assumed that $x_n \in K_{n+1} \setminus K_n$. (If necessary, pass to suitable subsequences of (x_n) and (K_n).) We need the following lemma which is an easy consequence of the Hahn-Banach theorem:

LEMMA: *Let* V *be a bounded closed convex subset of a Banach space* E *and let* $x \in E \setminus V$. *There exist* $\mu \in E'$ *and* $\alpha \in \mathbb{C}$ *such that*

$$|\mu(x) + \alpha| > \|\mu + \alpha\|_V.$$

So, because of i), one cand find $\mu_n \in E'$ and $\alpha_n \in \mathbb{C}$ such that

$\ell_n = \mu_n + \alpha_n \in \mathcal{H}(E)$ satisfies $|\ell_n(x_n)| > \|\ell_n\|_{K_n}$ for all $n \in \mathbb{N}$. By a suitable multiplication and by an induction argument one gets a sequence (f_n) of analytic functions $f_n \in \mathcal{H}(E)$ with

iii) $\|f_n\|_{K_n} < 2^{-n}$

and

iv) $|f_n(x_n)| \geqslant n + 1 + \sum_{j=1}^{n-1} |f_j(x_n)|$

for all $n \in \mathbb{N}$. From ii) and iii) it follows that $\sum f_n = f$ is an analytic function on E which satisfies $|f(x_n)| \geqslant n$ according to iv). Therefore $\|f\|_B = \infty$, and B is not bounding which completes the proof.

DEFINITION: A Banach space E will be called SP-*space* if for every separable subspace $E_o \subset E$ there exists a separable direct subspace E_1 of E such that $E_o \subset E_1$.

THEOREM: *In a* SP-*space* E *every bounding set is relatively compact.*

PROOF: Let B and (x_n) be as in the foregoing proof. There exists a continuous linear map $p: E \longrightarrow E$ such that $E_1 = p(E)$ is separable and contains (x_n). According to the proposition at least one holomorphic function g on E_1 is unbounded on (x_n). The trivial extension $f = g \circ p$ of g satisfies $\|f\|_B = \infty$.

A result of AMIR-LINDENSTRAUSS [2] gives a sufficient criterion for E to be a SP-space:

DEFINITION: E is called *weakly compact generated* if there exists a weakly compact set $K \subset E$ such that the linear hull of K is dense in E.

EXAMPLE: Every reflexive Banach space is weakly compact generated since the closed unit ball is weakly compact.

PROPOSITION [2]: *Every compact generated Banach space is a* SP-*space.*

The results of DINEEN and HIRSCHOWITZ mentioned in the introduction are the following: HIRSCHOWITZ proved in [9] that the above theorem is true for all Banach spaces E which can be embedded in a space $\mathcal{C}(T)$ of continuous functions on a sequentially compact space T: Each bounding set in E is relatively compact. This result includes that of DINEEN [6]: if E is a Banach space such that the closed unit ball B' of E' is weakly sequentially compact, then the bounding subsets of E are relatively compact. For such a space is embeddable in $\mathcal{C}(B')$.

An important example is given by DINEEN in [7]: In ℓ^∞ the non compact set of unit vectors $\{(\delta_{ij})_{j\in\mathbb{N}} \mid i \in \mathbb{N}\}$ is bounding. But the bounding sets in ℓ^∞ are at least nowhere dense in ℓ^∞ [6]. Whether this is true for all infinite dimensional Banach spaces is an open question.

2. ADMISSIBLE COVERINGS AND REGULAR CLASSES

Many of the classical methods of complex analysis in finite dimension use the fact that $\mathcal{H}(X)$ is a Fréchet algebra or that there are nice countable coverings of the domain X consisting of compact sets. In the infinite dimensional case $\mathcal{H}(X)$ is neither metrizable nor barrelled [1] and no such coverings exist. But analogous to the proof of the proposition in the first section one may consider subalgebras $A_{\mathcal{U}}$

of $\mathcal{H}(X)$ which are induced by admissible coverings $\mathcal{U}$ of X [13]. Relative to such algebras $A_{\mathcal{U}}$ one can define $A_{\mathcal{U}}$-convexity and prove characterizations similar to 1°-3° and (*).

DEFINITION: Let X be a domain in the Banach space E. An open covering $\mathcal{U}$ of X will be called *admissible* if

i) for every $U \in \mathcal{U}$ there exist $s > 0$ and $V \in \mathcal{U}$ such that $d_X(U) > s$ and $U_s \subset V$, where U_s is defined by $U_s = \bigcup\{B(x,s) \mid x \in U\}$;

ii) $\mathcal{U}$ is closed under finite unions

and

iii) every $U \in \mathcal{U}$ is bounded.

For an admissible covering $\mathcal{U}$ of X the algebra $A_{\mathcal{U}}$ is defined by

$$A_{\mathcal{U}} = \{f \in \mathcal{H}(X) \mid \|f\|_U < \infty \text{ for all } U \in \mathcal{U}\}.$$

$A_{\mathcal{U}}$ is a locally convex complete Hausdorff algebra with respect to the topology of uniform convergence on all $U \in \mathcal{U}$. $A_{\mathcal{U}}$ is a Fréchet algebra if $\mathcal{U}$ is countable. Because of iii) $A_{\mathcal{U}}$ contains all restrictions of continuous linear functions $\mu \in E'$. Further, $A_{\mathcal{U}}$ is a regular class:

DEFINITION: $A \subset \mathcal{H}(X)$ is called a *regular class* of analytic functions [3, p.121] if

i) $f^n \in A$ for all $f \in A$ and $n \in \mathbb{N}$,

ii) $(n!)^{-1}\, \hat{d}^n_a f \in A$ for all $f \in A$, $n \in \mathbb{N}$ and $a \in E$, where $\hat{d}^n_a f$ is defined by $\hat{d}^n_a f\colon X \ni x \longmapsto \hat{d}^n f(x)\cdot a \in \mathbb{C}$ when

$$f(x + a) = \sum (n!)^{-1} \hat{d}^n f(x) \cdot a$$

(see [12, p.17]) denotes the power series expansion of f at x,

iii) $\rho_A(x) = \inf\{\rho_f(x) \mid f \in A\} > 0$ for all $x \in X$, where the radius of convergence ρ_f of f is defined by

$$\rho_f(x) = \sup\{r \mid \sum \| (n!)^{-1} \hat{d}^n f(x) \| r^n < \infty\}.$$

The regularity of $A_{\mathcal{U}}$ follows easily from the Cauchy inequalities [12, p.22].

EXAMPLES: 1º - Let $f: X \longrightarrow F$ be an analytic map with values in a Banach space F. For $n \in \mathbb{N}$ define

$$U_n(f) = \{x \in X \mid d_X(x) > 2^{-n},\ \|x\| < n \text{ and } \|f\|_{B(x,2^{-n})} < n\}.$$

Then $\mathcal{U}(f) = (\overset{\circ}{\widehat{U_n(f)}})_{n \in \mathbb{N}}$ is an admissible and countable covering of X such that $\nu \circ f \in A_{\mathcal{U}(f)}$ for all $\nu \in F'$. It follows that

$$\mathcal{H}(X) = \bigcup \{A_{\mathcal{U}} \mid \mathcal{U} \in \omega(X)\},$$

if $\omega(X)$ denotes the set of all countable admissible coverings of S.

2º - The algebras $\mathcal{F}_\tau$ of COEURÉ [4, p.398] are examples of algebras $A_{\mathcal{U}}$ in the separable case. Therefore the following results are partially contained in [11].

3º - $A_{\mathcal{U}} = \mathcal{H}(X)$ if $\dim E < \infty$. Since $\rho_{\mathcal{H}(E)} \equiv 0$ is equivalent to the condition that every bounding set in E is nowhere dense, the results of the first chapter show that $\mathcal{H}(X)$ is in general not a regular class.

PROPOSITION: *Let* $\mathcal{U} \in \omega(X)$ *and* $A = A_{\mathcal{U}}$. *Then the following conditions are equivalent:*

1° - $d_X(\hat{A}(U)) > 0$ *for all* $U \in \mathcal{U}$, *where* $\hat{A}(U)$ *is defined by*

$$\hat{A}(U) = \{x \in X \mid |f(x)| \leqslant \|f\|_U \text{ for all } f \in A\}.$$

2° - *For all sequences* (x_n) *in* X *with* $\lim d_X(x_n) = 0$ *there exists* $f \in A$ *such that* $\sup\{|f(x_n)| \mid n \in \mathbb{N}\} = \infty$.

3° - $\rho_A \leqslant d_X$.

4° - *For all countable subsets* D *of* X *the set*

$$R(D) = \{f \in A \mid \text{for all } x \in D : \rho_f(x) \leqslant d_X(x)\}$$

is of second category in the Fréchet space A.

PROOF: 1° $\Longrightarrow$ 2°: As in the proof of the proposition in the first section we can assume that $x_n \in \hat{A}(U_{n+1}) \setminus \hat{A}(U_n)$, where $\mathcal{U} = (U_n)_{n \in \mathbb{N}}$. But then there exists a sequence (f_n) in A such that $\|f_n\|_{U_n} < 2^{-n}$ and

$$|f_n(x_n)| \geqslant n + 1 + \sum_{j=1}^{n-1} |f_j(x_n)|.$$

It follows that $\sum f_n = f \in A$ satisfies $|f(x_n)| \geqslant n$.

2° $\Longrightarrow$ 3°: Since $d_X(B(x, d_X(x))) = 0$ for a point $x \in X$ one can find a sequence (x_n) in $B(x, d_X(x))$ such that $d_X(x_n) \longrightarrow 0$. There exists $f \in A$ such that $\|f\|_{B(x,d_X(x))} = \infty$. Now $\rho_f(x) \leqslant d_X(x)$ follows from the Cauchy inequalities.

3° $\Longrightarrow$ 4°: Define

$$S^k(x) = \{f \in A \mid \rho_f(x) \geqslant d_X(x) + 2^{-k}\}$$

for $x \in D$. Each $f \in S^k(x)$ can be extended to an analytic function *f* on the union $\mathcal{X}$ of X and $B(x, d_X(x)+2^{-k})$ which may be a non schlicht domain if the analytic continuation is not unique. Let $\mathcal{V}$ be the admissible covering of $\mathcal{X}$ which consists of all finite unions of the sets $U \in \mathcal{U}$ and the balls $B(x, (1-2^{-m})d_X(x)+2^{-k})$, $m \in \mathbb{N}$. $A_{\mathcal{V}}$ is a Fréchet space and the restriction mapping

$$j^*: A_{\mathcal{V}} \ni g \longmapsto g|X \in A_{\mathcal{U}}$$

is linear continuous with the image $j^*(A_{\mathcal{V}}) = S^k(x)$. $S^k(x) \neq A_{\mathcal{U}}$ follows from 3°; hence, $S^k(x)$ is meager by the theorem of Banach and

$$R(D) = A \setminus \bigcup \{S^k(x) \mid k \in \mathbb{N},\ x \in D\}$$

is of second category, since A is a Baire space.

4° $\Longrightarrow$ 1°: Let $U \in \mathcal{U}$. There exist per definition $s > 0$ and $V \in \mathcal{U}$ such that $d_X(U) > s$ and $U_s \subset V$. For all $x \in \hat{A}(U)$ and $f \in A$ the Cauchy inequalities imply

$$\|(n!)^{-1} \hat{d}^n f(x)\| \leq s^{-n} \|f\|_{U_s} < s^{-n} \|f\|_V$$

and therefore $\rho_f(x) \geq s$. $d_X(\hat{A}(U)) \geq s > 0$ follows now from 4°.

DEFINITION: X will be called $A_{\mathcal{U}}$-*convex* if one of the above conditions 1°-4° is satisfied.

To illustrate the consequences of the foregoing proposition in the sense of (*) the notion of a domain spread over Banach spaces is needed. A domain X spread over the Banach space E is a connected Hausdorff space together with a local homeomorphism $p: X \longrightarrow E$. All the concepts above extend in a natural way to arbitrary domains and in particular the last proposition remains true [13].

DEFINITION: Let X, $\mathcal{X}$ be domains spread over E with projections p, $\mathcal{p}$. Let A be a family of analytic mappings defined on X. $j: X \longrightarrow \mathcal{X}$ is called a *simultaneous analytic continuation* (*s.a.c.*) of A if j is continuous and satisfies $p = \mathcal{p} \circ j$ and if there exists for every $f \in A$ an analytic map $\mathcal{f}$ on $\mathcal{X}$ such that $f = \mathcal{f} \circ j$.

Such a *s.a.c.* is called *maximal* if for all *s.a.c.* $j': X \longrightarrow X'$ of A there exists a unique *s.a.c.* $\mathcal{j}': X' \longrightarrow \mathcal{X}$ of $A' = \{f' \mid f' \circ j' \in A\}$ with $\mathcal{j} = \mathcal{j}' \circ j'$. The maximal *s.a.c.* of $\mathcal{H}(X)$ (which is unique up to isomorphisms [5], [10], [13]) is called the *envelope of holomorphy*.

X is called an *A-domain of holomorphy* if for all *s.a.c.* of A $j: X \longrightarrow \mathcal{X}$ the map j is an isomorphism of domains. X is a *domain of holomorphy* if X is a $\mathcal{H}(X)$-domain of holomorphy. X is a *domain of existence* of the analytic map $f: X \longrightarrow F$ if X is a $\{f\}$-domain of holomorphy.

THEOREM: *Let* $\mathcal{U}$ *be an admissible and countable covering of the domain* X *spread over* E. *Then* X *is an* $A_{\mathcal{U}}$*-domain of holomorphy if and only if* X *is* $A_{\mathcal{U}}$*-convex and* $A_{\mathcal{U}}$ *separates points.*

PROOF: If $j: X \longrightarrow \mathcal{X}$ is a *s.a.c.* of $A_{\mathcal{U}}$ and X is $A_{\mathcal{U}}$-convex, then j is onto by property 2º of the foregoing proposition. If $A_{\mathcal{U}}$ separates the points of X, j must be also injective. Hence, j is an isomorphism. Conversely, let X be an $A_{\mathcal{U}}$-domain of holomorphy. It follows that $\rho_{A_{\mathcal{U}}} \leqslant d_X$, otherwise one could find a proper *s.a.c.* of $A_{\mathcal{U}}$. Let $x \sim y$ if and only if $f(x) = f(y)$ for all $f \in A_{\mathcal{U}}$. Then $X \longrightarrow X/\sim$ is a *s.a.c.* which shows that $A_{\mathcal{U}}$ separates points.

REMARK: $A_{\mathcal{U}}$ separates the points of X if $\mathcal{U}$ is an admissible covering of a domain $X \subset E$.

In the separable case one gets the stronger result:

THEOREM: *Let* X *be a domain spread over the separable Banach space* E. *Then the following properties are equivalent:*

1º - X *is a domain of existence.*

2º - *There exists a regular class* A *on* X *such that* X *is an* A-*domain of holomorphy.*

3º - *There exists an admissible covering* $\mathcal{U}$ *of* X *such that* X *is* $A_{\mathcal{U}}$-*convex and* $A_{\mathcal{U}}$ *separates points.*

4º - X *is the domain of existence of an analytic function* $f: X \longrightarrow \mathbf{C}$.

Sketch of the proof: 1º $\Longrightarrow$ 2º: Let X be the domain of existence of the map $f: X \longrightarrow F$ and define $\mathcal{U} = \mathcal{U}(f)$ as in the example 1º. Then $A = A_{\mathcal{U}}$ separates points and X is A-convex. Hence, X is an A-domain of holomorphy according to the above theorem.

2º $\Longrightarrow$ 3º: There exists a countable admissible covering $\mathcal{U}$ of X such that $A \subset A_{\mathcal{U}}$ and, of course, X is an $A_{\mathcal{U}}$-domain of holomorphy. As a consequence X is $A_{\mathcal{U}}$-convex and $A_{\mathcal{U}}$ separates points.

3º $\Longrightarrow$ 4º: It can be assumed that $\mathcal{U}$ is countable. Let $Q \subset E$ be dense and countable then $D = p^{-1}(Q)$ is countable and dense in X.

$$G = \{(x,y) \in X \times X \mid x \neq y, \quad px = py \in Q\}$$

is also countable. Since $A_{\mathcal{U}}$ separates the points of X the set

$$T(G) = \{f \in A_{\mathcal{U}} \mid f(x) \neq f(y) \quad \text{for all} \quad (x,y) \in G\}$$

is of second category in $A_{\mathcal{U}}$. Since X is $A_{\mathcal{U}}$-convex the set $R(D)$

is of second category (condition 4º of the proposition). Because of the continuity of ρ_f and d_X all $f \in R(D)$ satisfy $\rho_f \leqslant d_X$. Now $H = T(G) \cap R(D)$ is of second category, in particular not empty, and X is the domain of existence of each $f \in H$.

4º $\Longrightarrow$ 1º is evident.

The following proposition plays a key role for the construction of the envelope of holomorphy via the spectrum of $\mathcal{H}(X)$,

PROPOSITION: *Let* $j: X \longmapsto X$ *be s.a.c. of* $A_{\mathcal{U}}$, $\mathcal{U} \in \omega(X)$, *and let* $A = \{f \mid f \in A_{\mathcal{U}}\}$ *where* $f \in \mathcal{H}(X)$ *denotes the extension of* $f \in A_{\mathcal{U}}$. *Then*

1º - $\hat{x}: A_{\mathcal{U}} \ni f \longmapsto f(x) \in \mathbf{C}$ *is continuous for all* $x \in X$.

2º - *There exists* $\mathcal{V} \in \omega(X)$ *such that* $A = A_{\mathcal{V}}$ *and* $j^*: A_{\mathcal{V}} \longrightarrow A_{\mathcal{U}}$ *is an isomorphism of Fréchet spaces.*

Sketch of the proof: 1º: The set $W = \{x \in X \mid \hat{x} \text{ is continuous}\}$ is closed: Let $x_n \in W$ and $x_n \longrightarrow x \in X$. $(\hat{x}_n) \subset (A_{\mathcal{U}})'$ is weakly bounded, hence equicontinuous ($A_{\mathcal{U}}$ is barrelled!). It follows that there exists $U \in \mathcal{U}$ such that $|\hat{x}_n(f)| \leqslant \|f\|_U$ for all $f \in A_{\mathcal{U}}$, and then $|\hat{x}(f)| \leqslant \|f\|_U$, i.e. $x \in W$. W is also open: Let $x \in W$ and $U \in \mathcal{U}$ such that $|\hat{x}(f)| \leqslant \|f\|_U$ for all $f \in A_{\mathcal{U}}$. There exist $V \in \mathcal{U}$ and $0 < s < d_X(x)$ such that $d_X(U) > s$ and $U_s \subset V$. The Cauchy inequalities imply $|\hat{y}(f)| \leqslant \|f\|_{U_s} \leqslant \|f\|_V$ for all $y \in B(x,s)$ and $f \in A_{\mathcal{U}}$. Hence, $B(x,s) \subset W$ and W being open and closed is the whole domain X.

2º. For $U \in \mathcal{U}$ and $n \in \mathbb{N}$ define $V_n(U)$ to be the interior of the set

$$\{x \in X \mid d_X(x) > 2^{-n},\ \|px\| < n \text{ and } B(x,2^{-n}) \subset \hat{A}(j(U))\}.$$

The family of finite unions of the sets $V_n(U)$, $n \in \mathbb{N}$, $U \in \mathcal{U}$, is an admissible covering $\mathcal{V}$ of X and satisfies 2º.

3. APPLICATIONS

As immediate applications of the last proposition in the second section one gets the following two propositions [13]:

PROPOSITION: *Let* $\mathcal{H}(X)_b$ *denote the vector space* $\mathcal{H}(X)$ *endowed with the inductive limit topology of the system* $(A_{\mathcal{U}} \mid \mathcal{U} \in \omega(X))$. *Then for each s.a.c.* $j: X \longrightarrow \mathfrak{X}$ *of* $\mathcal{H}(X)$ *the restriction map*

$$j^*: \mathcal{H}(\mathfrak{X})_b \longrightarrow \mathcal{H}(X)_b$$

is an isomorphism of topological vector spaces (see also [1], [4] and [10]).

PROPOSITION: *Let* $f: X \longrightarrow F$ *be an analytic map with values in the Banach space* F *and let* $j: X \longrightarrow \mathfrak{X}$ *be a s.a.c. of* $A_{\mathcal{U}(f)}$ (see ex. 1º). *Then* f *has an analytic continuation* $\mathfrak{f}: \mathfrak{X} \longrightarrow F$ *with* $f = \mathfrak{f} \circ j$.

COROLLARY: *Every analytic map* $f: X \longrightarrow F$ *can be extended analytically to the envelope of holomorphy of* X (see also [4] and [10]).

The maximal *s.a.c.* of a regular class $A_{\mathcal{U}} \subset \mathcal{H}(X)$, $\mathcal{U} \in \omega(X)$, can be constructed by taking a certain subset of the spectrum of A [13]. As an application one gets a construction of the envelope of holomorphy of X via the spectrum of $\mathcal{H}(X)$ (se [8, p.49] for the finite dimensional case). With the aid of this result the foregoing corollary remains true if one replaces F by an $A_{\mathcal{U}}$-domain of holomorphy, $\mathcal{U}$ admissible and countable.

A further application of the results on regular classes leads to a characterization of those normed separable spaces whose $\mathcal{O}$-completion is not the completion [5].

One can define admissible coverings and regular classes for domains spread over arbitrary locally convex Hausdorff spaces. Analogous results like in section 2. can be proved. The applications in this section remain true at least for the metrizable domains.

REFERENCES

[1] H.ALEXANDER, *Analytic Functions on Banach Spaces*. Thesis, University of California (1968).

[2] D. AMIR - J. LINDENSTRAUSS, *The Structure of Weakly Compact Sets in Banach Spaces*, Ann. of Math. 88(1968), 35-46.

[3] H. BEHNKE - P. THULLEN, Theorie der Funktionen mehrer komplexer Veränderlichen, Erg. der Math. 51(1970), First edition 1932.

[4] G. COEURÉ, *Fonctions plurisousharmoniques sur les espaces vectoriels topologiques et applications a l'étude des fonctions analytiques*, Ann. Inst. Fourier 20 (1970), 361-432.

[5] G. COEURÉ, Forthcoming book on complex manifolds spread over normed spaces, To appear at North-Holland.

[6] S. DINEEN, *Unbounded Holomorphic Functions on a Banach Space*, J. London Math. Soc. **4**(1972), 461-465.

[7] S. DINEEN, *Bounding Subsets of a Banach Space*, Math. Ann. **192**(1971), 61-70.

[8] R.C. GUNNING - H. ROSSI, Analytic Functions of Several Complex Variables, Englewood Cliffs, N.J., Prentice-Hall (1965).

[9] A. HIRSCHOWITZ, *Bornologie des espaces de fonctions analytiques en dimension infinie*, Sém. LELONG 1970, Springer Lecture Notes, **205** (1971).

[10] A. HIRSCHOWITZ, *Prolongement analytique en dimension infinie*, C.R. Acad. Sci. **270**(1970), 1136-1137, And to appear in Ann. Inst. Fourier.

[11] M. MATOS, *Domains of τ-Holomorphy in a Separable Banach Space*, Math. Ann. **195**(1972), 273-278.

[12] L. NACHBIN, Topology on Spaces of Holomorphic Mappings, Erg. d. Math. **47**(1969).

[13] M. SCHOTTENLOHER, Analytische Fortsetzung in Banachräumen, Dissertation, Universität München (1970) and in parts in Math. Ann. **199**(1972), 313-336.

University of Munich
Department of Mathematics
8 Munich 2 Theresienstrasse 39
W.GERMANY

MODULES OF CONTINUOUS FUNCTIONS

by

João B. Prolla

Let X be a completely regular Hausdorff space and $V > 0$ a directed set of non-negative upper-semicontinuous functions on X. If E is a locally convex Hausdorff space, we denote by $C(X; E)$ (resp. $C(X)$) the set of all continuous mappings from X into E (resp. into $K = \mathbb{R}$ or $\mathbb{C}$). The vector subspace of all $f \in C(X; E)$ such that vf vanishes at infinity for every $v \in V$, is denoted by $CV_\infty(X; E)$ and will be equipped with the topology determined by the seminorms

$$f \longrightarrow \sup\{v(x)p(f(x)) \ ; \ x \in X\}$$

for $f \in CV_\infty(X;E)$, where $v \in V$ and p ranges over the set of continuous seminorms on E.

In what follows, $M \subset C(X)$ denotes a $C_b(X)$-module, i.e. M is a vector subspace of $C(X)$ closed under multiplication by *bounded* continuous functions. The following "partition of unity" result, due to Nachbin, will be needed. (For a proof see Nachbin [4], §23.)

LEMMA. *Let* $K \subset\!\!\!\complement X$ *be a compact subset such that* $K \subset Z(M)$. *If* $A_i \subset X$ $(i = 1,2,\dots,n)$ *is an open covering of* K, *there are* $g_i \in M$ *such that* $g_i = 0$ *outside of* A, *and* $g_i \geqslant 0$ *on* X $(i = 1,2,\dots,n)$, $\sum g_i = 1$ *on* K *and* $\sum g_i \leqslant 1$ *on* X.

NOTATION. $Z(M) = \{x \in X \ ; \ g(x) = 0 \ \text{for all} \ g \in M\}$.

Given a vector subspace $W \subset CV_\infty(X; E)$ which is an M-module, i.e. such that $MW \subset W$, the *weighted approximation problem* consists in finding a characterization of the closure of W in $CW_\infty(X; E)$. This is accomplished in the following theorem.

THEOREM 1. *Let* $W \subset CV_\infty(X; E)$ *be a vector subspace which is an* M-*module such that* $Z(M) \subset Z(W)$. *A function* $f \in CV_\infty(X; E)$ *belongs to the closure of* W *if, and only if, for every* $x \in X$, $f(x)$ *belongs to the closure of* $W(x) = \{w(x) ; w \in W\}$ *in* E.

REMARK. In many instances the hypothesis $Z(M) \subset Z(W)$ is trivially satisfied because $Z(M) = \emptyset$, i.e. for each $x \in X$ there is $g \in M$ such that $g(x) \neq 0$.

Proof of th. 1. The condition is obviously necessary. Conversely, let $f \in CV_\infty(X; E)$ be such that $f(x)$ belongs to the closure of $W(x)$ in X, for each $x \in E$. Let $v \in V$, $\varepsilon > 0$ and p a continuous seminorm on E be given. Then $K = \{x \in X ; v(x)p(f(x)) \geq \varepsilon\}$ is a compact subset of X. For each $t \in K$, there is $w_t \in W$ such that

$$v(t)p(f(t) - w_t(t)) < \varepsilon.$$

Since $t \in K$, $w_t(t) \neq 0$. Hence there is $g_t \in M$ such that $g_t(t) \neq 0$. Notice that the mapping $x \longmapsto v(x)p(f(x) - w_t(x))$ is upper-semicontinuous. Therefore, an open neighborhood U_t of t in X can be found such that for $x \in U_t$ we have $v(x)p(f(x) - w_t(x)) < \varepsilon$. By compactness of K, there are $t_1, \ldots, t_n \in K$ such that $A_i = U_{t_i}$ $(i = 1,2,\ldots,n)$ form an open covering of K. By our previous remark, $K \subset \complement Z(M)$, and we can apply the Lemma. Let $g_i \in M$ be such that $g_i = 0$ outside of A_i and $g_i \geq 0$ on X $(i = 1,2,\ldots,n)$, $\sum g_i = 1$ on K and $\sum g_i \leq 1$ on X. Let $h_i = w_{t_i}$ $(i = 1,2,\ldots,n)$. We claim that

(1) $$v(x)p(f(x) - \sum g_i(x)h_i(x)) < 3\varepsilon$$

for all $x \in X$. Indeed, if $x \in K$, then

$$v(x)p\left[f(x) - \sum g_i(x)h_i(x)\right] = v(x)p\left[\sum g_i(x)\ f(x) - h_i(x)\right]$$
$$\leq \sum g_i(x)v(x)p\left[f(x) - h_i(x)\right]$$

and (1) follows from

$$g_i(x)v(x)p\left[f(x) - h_i(x)\right] \leq g_i(x)\varepsilon,$$

valid for all $x \in X$ and $i = 1,2,\ldots,n$. If $x \notin K$, then

$$v(x)p\left[f(x) - \sum g_i(x)h_i(x)\right] \leq v(x)p\left[f(x)\right] + \sum g_i(x)v(x)p\left[h_i(x)\right]$$

and (1) follows from

$$g_i(x)v(x)p\left[h_i(x)\right] \leq g_i(x)\cdot 2\varepsilon,$$

valid for $x \in X$ and $i = 1,2,\ldots,n$. Since $\sum g_i h_i$ belongs to $MW \subset W$, it follows from (1) that f belongs to the closure of W in $CV_\infty(X;E)$.

COROLLARY 1. *Under the hypothesis of Theorem 1 suppose that* $W(x)$ *is dense in* E *for each* $x \in X$. *Then* W *is dense in* $CV_\infty(X; E)$.

COROLLARY 2. *Let* $W \subset CV_\infty(X)$ *be an M-module, dense in* $CV_\infty(X)$. *Then* $Z(M) \subset Z(W)$ *implies that* $W \otimes E$ *is dense in* $CV_\infty(X; E)$. *In particular,* $CV_\infty(X) \otimes E$ *is dense in* $CV_\infty(X; E)$.

COROLLARY 3. *Under the hypothesis of Theorem 1 let* $W_o = W \circ E'$ *and suppose that* $W_o \otimes E \subset W$. *Then* W_o *total in* $CV_\infty(X)$ *implies that* W *is dense in* $CV_\infty(X; E)$.

THEOREM 2. *Let* $E = K$ $(K = \mathbb{R}$ *or* $\mathbb{C})$, *and* W *be as in Theorem 1. Then* $f \in CV_\infty(X)$ *belongs to the closure of* W *if, and only if,* f *vanishes on* $Z(W)$.

PROOF. The condition is obviously necessary. Suppose now that $f \in CV_\infty(X)$ vanishes on $Z(W)$. Let $x \in X$. If $f(x) = 0$, then obviously $f(x) \in W(x)$. If $f(x) \neq 0$, then $x \notin Z(W)$ and there exists $w \in W$ such that $w(x) \neq 0$. Consider $g = [f(x)/w(x)]\, w \in W$. Then $f(x) = g(x) \in W(x)$. By Theorem 1, f belongs to the closure of W.

REMARK. In many instances M and $CV_\infty(X)$ satisfy the following properties:

(1) $Z(M) = \emptyset$;

(2) given $x \in X$ and $F \subset X$ a closed subset not containing x, there is $f \in CV_\infty(X)$ such that $f(x) \neq 0$ and $f(t) = 0$ for all $t \in F$.

THEOREM 3. *Let* M *and* $CV_\infty(X)$ *satisfy conditions (1) and (2) above. Then for every closed* M*-module* $W \subset CV_\infty(X)$ *there is a unique closed subset* $N_W \subset X$ *such that*

$$W = \{f \in CV_\infty(X) \;;\; f(x) = 0 \text{ for all } x \in N_W\}.$$

PROOF. Since W is closed, Theorem 2 implies that

$$W = \{f \in CV_\infty(X) \;;\; f(x) = 0 \text{ for all } x \in Z(W)\}.$$

Let $N_W \subset X$ be any other closed subset of X such that

$$W = \{f \in CV_\infty(X) \;;\; f(x) = 0 \text{ for all } x \in N_W\}.$$

Obviously, $N_W \subset Z(W)$. Suppose that the inclusion is proper; i.e. there exists $x \in Z(W)$ such that $x \notin N_W$. By condition (2) there exists $f \in CV_\infty(X)$ such that $f(x) \neq 0$ and $f(t) = 0$ for all $t \in N_W$. Therefore $f \notin W$ but vanishes on N_W, a contradiction. Hence $N_W = Z(W)$.

The above theorems 2 and 3 provide a one-to-one correspondence between the closed ideals of several algebras of continuous functions and the closed subsets of X, when X is locally compact. Indeed, let

$A_1 = C(X)$, with the compact-open topology,

$A_2 = C_b(X)$, with the strict topology β,

$A_3 = C_\infty(X)$, with the uniform topology.

Let $I \subset A_i$ $(i = 1,2,3)$ be a closed ideal. Then there exists a unique closed subset $N \subset X$ such that

$$I = \{f \in A_i \; ; \; f(x) = 0 \text{ for all } x \in N\} \quad (i = 1,2,3).$$

Indeed, $A_i = (CV_i)_\infty(X)$, $i = 1,2,3$, where

V_1 = set of characteristic functions of compact subsets of X;

$V_2 = C_\infty^+(X)$, the set of all positive continuous functions on X that vanish at ∞;

$V_3 = \{1\}$, where $1(x) = 1$ for all $x \in X$.

Then $V_i > 0$ $(i = 1,2,3)$ and to apply Theorem 3 consider $M = K(X)$, the set of all continuous functions on X with compact support.

TERMINOLOGY. Given $M \subset C(X)$, a $C_b(X)$-module, and $W \subset CV_\infty(X; E)$ an M-module, we say that (M, W) is an admissible pair if $Z(M) \subset Z(W)$ i. e. if they satisfy the hypothesis of Theorem 1.

Given two Nachbin spaces $(CV_1)_\infty(X_1; E)$ and $(CV_2)_\infty(X_2; E)$, let $X = X_1 \times X_2$ and let V be the set of maps

$$(x_1, x_2) \in X \longmapsto v_1(x_1)v_2(x_2)$$

for each choice of $v_i \in V_i$ $(i = 1,2)$. If $W_i \subset (CV_i)_\infty(X_i; E)$ are vector subspaces, the *slice product* $W_1 \# W_2$ is the vector subspace of all $f \in CV_\infty(X; E)$ such that the mappings

$$x \in X_1 \longmapsto f(x, a_2) \quad \text{and} \quad x \in X_2 \longmapsto f(a_1, x)$$

belong respectively to W_1 and W_2, for each choice of $(a_1,a_2) \in X$. If (M_1,W_1) and (M_2,W_2) are admissible pairs, then $(M_1 \# M_2 , W_1 \# W_2)$ is an admissible pair.

THEOREM 4. $CV_\infty(X) = (CV_1)_\infty(X_1) \# (CV_2)_\infty(X_2)$.

COROLLARY. $(CV_1)_\infty(X_1) \otimes (CV_2)_\infty(X_2)$ *is dense in* $CV_\infty(X)$.

For proofs see Prolla [5].

REMARKS. Other properties of slice products of Nachbin spaces can be found in Bierstedt [1]. For slice products of function algebras, see Birtel [2] and Eifler [3]. For generalizations and more details of the results presented here, see Prolla [5].

BIBLIOGRAPHY

[1] K.-D. BIERSTEDT, Gewichtete Räume Stetiger Vektorwertiger Funktionen und das Injektive Tensorprodukt, Doctoral Dissertation, Johannes Gutenberg Universität, Mainz, 1970.

[2] F.T.BIRTEL, *Slice algebras of bounded analytic functions*, Math. Scand. 21(1967), 54-60.

[3] L.EIFLER, *The slice product of function algebras*, Proc. Amer. Math. Soc. 23(1969), 559-564.

[4] L.NACHBIN, Elements of Approximation Theory, D. van Nostrand Co., Princeton, 1967.

[5] J.B.PROLLA, *Weighted approximation and slice products of modules of continuous functions*, Ann. Scuola Norm. Sup. Pisa, to appear.

Universidade Federal do Rio de Janeiro
Instituto de Matemática
Rio de Janeiro, GB, ZC-32
BRASIL

WAVE FRONT-SETS

by
Karl G. Andersson

This lecture is about a refinement of the notion of singular support of a distribution. If u is a distribution in the open set $\Omega \subset \mathbb{R}^n$, the singular support of u, S.S.u, is the smallest, relatively closed subset of Ω such that $u \in C^\infty(\Omega \setminus \text{S.S. } u)$. The singular support may be described by means of Fourier transforms as follows.

LEMMA 1. *$x_0 \notin$ S.S.u if and only if there is a function $\phi \in C_0^\infty(\Omega)$ such that $\phi = 1$ in a neighborhood of x_0 and*

$$|\widehat{\phi u}(\xi)| < C_N(1 + |\xi|)^{-N}, \quad N = 1,2,\ldots . \tag{1}$$

Here $\widehat{\phi u}$ denotes the Fourier transform of ϕu.

PROOF. Suppose that $x_0 \notin$ S.S.u. Then $\phi u \in C_0^\infty(\Omega)$ if ϕ vanishes outside a small neighborhood of x_0. Thus

$$\xi^\alpha \widehat{\phi u}(\xi) = (-1)^{|\alpha|}\int (D_x^\alpha e^{-i<x,\xi>})(\phi u)(x)\,dx = \tag{2}$$

$$= \int e^{-i<x,\xi>}(D_x^\alpha \phi u)(x)\,dx.$$

Here $\alpha = (\alpha_1,\ldots,\alpha_n)$ is a multi-index, $|\alpha| = \alpha_1 + \ldots + \alpha_n$, $\xi^\alpha = \xi_1^{\alpha_1} \ldots \xi_n^{\alpha_n}$, $D_x^\alpha = D_{x_1}^{\alpha_1} \ldots D_{x_n}^{\alpha_n}$ and $D_{x_k} = \frac{1}{i}\frac{\partial}{\partial x_k}$. From (2) we get

$$|\xi^\alpha||\widehat{\phi u}(\xi)| \leq C_\alpha$$

and (1) follows.

Assume now that (1) is satisfied. The Fourier inversion formula gives that

$$(\phi u)(x) = (2\pi)^{-n} \int e^{i<x,\xi>} (\hat{\phi u})(\xi) d\xi$$

and it follows from (1) that all derivatives of u are continuous and may be computed by differentiating under the integral sign. Since $\phi u = u$ in a neighborhood of x_0, this proves that $x_0 \notin$ S.S.u.

Even if $x_0 \in$ S.S.u it may happen that (1) is satisfied in certain ξ-directions. We note $\mathbb{R}^n \setminus \{0\}$ by $\dot{\mathbb{R}}^n$ and make the following definition.

DEFINITION 1. Suppose that $u \in D'(\Omega)$ and that $(x_0, \xi_0) \in \Omega \times \dot{\mathbb{R}}^n$. We say that $(x_0, \xi_0) \notin WF(u)$ if and only if there is a function $\phi \in C_0^\infty(\Omega)$, with $\phi = 1$ in a neighborhood of x_0, and a conic neighborhood Γ of ξ_0 such that

(3) $$|\hat{\phi u}(\xi)| \leq C_N (1 + |\xi|)^{-N}, \quad n = 1,2,\ldots, \quad \text{when} \quad \xi \in \Gamma.$$

The set $WF(u)$ is called the *wave front set* of u.

$WF(u)$ is a subset of $\Omega \times \dot{\mathbb{R}}^n$ or, if one wants to emphasize the behavior under coordinate transformations, $\dot{T}^*(\Omega)$, where the dot indicates that the zero-section is removed. More precisely it can be proved that if $y = \kappa(x)$ is a C^∞ change of coordinates in Ω and β denotes the map on $\Omega \times \dot{\mathbb{R}}^n$ defined by

$$\beta(x,\xi) = (\kappa(x), ({}^t\kappa'(x))^{-1}(\xi)),$$

then

$$WF(u \circ \kappa^{-1}) = \beta(WF(u)).$$

If we denote by π the natural projection $T^*(\Omega) \longrightarrow \Omega$, it follows easily from Lemma 1 that

(4) $$\pi(WF(u)) = S.S.u.$$

It is also useful to make a similar refinement of the analytic singular support, A.S. u, which is the smallest relatively closed subset of Ω such that u is analytic in $\Omega \setminus A.S.u$. That u is analytic at $x_0 \in \Omega$ means that u has a convergent power series expansion around x_0 or, in other words, that $u \in C^\infty$ in a neighborhood of x_0 and

(5) $$|D^\alpha u(x)| \leq C^{|\alpha|+1} |\alpha|!, \text{ when } x \text{ is close to } x_0.$$

In view of Stirling's formula the estimates (5) are equivalent to estimates of the form

(5') $$|D^\alpha u(x)| \leq C_1^{|\alpha|+1} |\alpha|^{|\alpha|}.$$

To give a characterization of A.S.u, corresponding to Lemma 1, one has to be a little bit careful, since there are no non-trivial analytic functions with compact support. The following lemma provides the necessary tool.

LEMMA 2. *Let* K *be a compact subset of the open set* $\Omega \subset \mathbb{R}^n$. *Then there is a constant* C *and a sequence* $\phi_N \in C_0^\infty(\Omega)$ *such that* $\phi_N = 1$ *on* K *and*

(6) $$\sup_{x \in \Omega} |D^\alpha \phi_N(x)| \leq C^{|\alpha|+1} N^{|\alpha|}, \quad |\alpha| \leq N.$$

PROOF. Let ψ be a non-negative function in $C_o^\infty(\mathbb{R}^n)$ such that $\psi(x) = 0$ when $|x| \geqslant \varepsilon$, $\int \psi(x)dx = 1$ and put $\psi_{(N)}(x) = N^n\psi(Nx)$. If the function $\chi \in C_o^\infty(\Omega)$ equals 1 on a neighborhood of K and ε is small enough, we can define ϕ_N as $\chi*\psi_{(N)}*\dots*\psi_{(N)}$, where the convolution contains the function $\psi_{(N)}$ N times. Since $\psi_{(N)}(x) = 0$ if $|x| \geqslant \frac{\varepsilon}{N}$ we have $\phi_N \in C_o^\infty(\Omega)$ if ε is small enough. We can also assume that $\chi = 1$ on $K - \{x, |x| \leqslant \varepsilon\}$ and therefore

$$\phi_N(x) = (\int \psi_{(N)}(x)dx)^N = (\int \psi(x)dx)^N = 1, \quad \text{when} \quad x \in K.$$

Finally, if $|\alpha| = k \leqslant N$, we have

$$D^\alpha \phi_N = \chi * D^{\alpha^1}\psi_{(N)} * \dots * D^{\alpha^k}\psi_{(N)} * \psi_{(N)} * \dots * \psi_{(N)} ,$$

where $|\alpha^i| = 1$, $i = 1,\dots,k$. Since

$$|(f * g)(x)| \leqslant (\int |f(x)|dx)(\int |g(x)|dx),$$

when $f,g \in C_o^\infty(\mathbb{R}^n)$, and $D^{\alpha^i}\psi_{(N)}(x) = N(D^{\alpha^1}\psi)_{(N)}(x)$ we get, with

$$C = \max_i \int |D_{x_i}\psi|dx,$$

that

$$|D^\alpha\phi_N(x)| \leqslant (\int |\chi(x)|dx)(CN)^k.$$

This proves (6).

The following lemma now follows in the same way as Lemma 1.

LEMMA 1'. $x_0 \notin$ A.S.u *if and only if there is a sequence* $\phi_N \in C_0^\infty(\Omega)$, *satisfying (6), such that* $\phi_N = 1$ *in a fixed neighborhood of* x_0 *and*

$$(7) \qquad |\widehat{\phi_N u}(\xi)| \leqslant C^{N+1} N^N (1 + |\xi|)^{-N}, \quad N = 1,2,\dots .$$

In analogy with Definition 1 we have

DEFINITION 1'. Suppose that $u \in D'(\Omega)$ and that $(x_0,\xi_0) \in \Omega \times \dot{\mathbb{R}}^n$. We say that $(x_0,\xi_0) \notin WF_A(u)$ if and only if there is a sequence $\phi_N \in C_0^\infty(\Omega)$, satisfying (6) with $\phi_N = 1$ in a fixed neighborhood of x_0, and a conic neighborhood Γ of ξ_0 such that

$$(8) \qquad |\widehat{\phi_N u}(\xi)| \leqslant C^{N+1} N^N (1 + |\xi|)^{-N}, \quad N = 1,2,\dots, \quad \text{when} \quad \xi \in \Gamma.$$

The set $WF_A(u)$ is called the *analytic wave front set* of u.

Again $WF_A(u)$ behaves like a subset of $T^*(\Omega)$ under analytic coordinate transformations in Ω and we have

$$(4') \qquad \pi(WF_A(u)) = \text{A.S.u.}$$

REMARK. The first one to observe the advantages of studying subsets of $T^*(\Omega)$, like WF(u), instead of subsets of Ω, like S.S.u, was M. Sato, who at the conference on Functional Analysis in Tokyo 1969 introduced some concepts for hyperfunctions similar to wave front sets. The definitions I have given here are due to L. Hörmander.

A reason for the rapid success of wave front sets in the study of linear differential equations is that many important invariants of the equations, such as characteristic surfaces, are most naturally, considered as subsets of $T^*(\Omega)$ rather than Ω.

If $P(x,D) = \sum_{|\alpha|\leq m} a_\alpha(x)D^\alpha$ is a linear differential operator with C^∞ coefficients a_α and $P_m(x,D) = \sum_{|\alpha|=m} a_\alpha(x)D^\alpha$ its principal part, we have

THEOREM 1. *If* $P(x,D)u \in C^\infty(\Omega)$, *then*

$$WF(u) \subset Z(P_m) = \{(x,\xi) \in \Omega \times \dot{\mathbb{R}}^n \; ; \; P_m(x,\xi) = 0\}.$$

A special case is the classical

COROLLARY 1. *If* $P(x,D)$ *is elliptic, i.e.* $P(x,\xi) \neq 0$ *when* $(x,\xi) \in \Omega \times \dot{\mathbb{R}}^n$, *and* $P(x,D)u \in C^\infty(\Omega)$ *then* $u \in C^\infty(\Omega)$.

For operators with analytic coefficients we have

THEOREM 1'. *If* $P(x,D)u$ *is analytic, then*

$$WF_A(u) \subset Z(P_m).$$

COROLLARY 1'. *If* $P(x,D)$ *is elliptic and* $P(x,D)$ *is analytic then* u *is analytic.*

For non-elliptic operators more precise results can be obtained by considering the so called bicharacteristics of $P(x,D)$. Suppose that $P(x,D)$ has real coefficients and put

$$H_{P_m} = \sum_j \left(\frac{\partial P_m(x,\xi)}{\partial \xi_j} \frac{\partial}{\partial x_j} - \frac{\partial P_m(x,\xi)}{\partial x_j} \frac{\partial}{\partial \xi_j} \right).$$

The vector field H_{P_m} is tangent to $Z(P_m)$ and if H_{P_m} is non-degenerate we get a one dimensional foliation of $Z(P_m)$. The integral curves in $Z(P_m)$ corresponding to H_{P_m} are called *bicharacteristic strips.*

THEOREM 2. *Suppose that* $P_m(x,D)$ *has real coefficients and that* H_{P_m} *is non-degenerate. If* $P(x,D)u \in C^\infty(\Omega)$ *then* $WF(u)$ *is invariant under the flow generated by* H_{P_m} *in* $Z(P_m)$.

Results of this kind about the "propagation of singularities" are closely connected with the problem of the surjectivity of the mapping $P(x,D): D'(\Omega)/C^\infty(\Omega) \longrightarrow D'(\Omega)/C^\infty(\Omega)$. To establish the surjectivity of the mapping $P(x,D): C^\infty(\Omega) \longrightarrow C^\infty(\Omega)$ one has to study the "propagation of zeros", i.e. prove uniqueness theorems. Also here wave front sets are useful. The following result, due to Hörmander, may be considered as a generalization of the principle of analytic continuation.

THEOREM 3. *Suppose that* Ω_0 *is an open subset of* Ω *with* C^1*-boundary* $\partial\Omega_0$. *Denote by* N_0 *a normal of* $\partial\Omega_0$ *at* $x_0 \in \Omega$ *and let* F *be a conic neighborhood of* $(x_0, \pm N_0)$, *i.e.* $F = U \times \Gamma$ *where* Γ *is conic. Then there is a neighborhood* Ω' *of* x_0 *such that any* u, *with* $WF_A(u) \cap F = \emptyset$, *which vanishes in* Ω_0 *must also vanish in* Ω'.

We shall prove this theorem under the extra assumption that u is continuous and $\partial\Omega_0$ is analytic in a neighborhood of x_0. The general result may be proved in the same way except for some technical complications.

Since $\partial\Omega_0$ is analytic and $WF_A(u)$ behaves like a subset of $T^*(\Omega)$ under analytic coordinate transformations in Ω, we may assume that $x_0 = 0$ and that Ω_0 is given by

$$x_n \leqslant |x'|^2 = x_1^2 + \ldots + x_{n-1}^2.$$

It will be sufficient to prove that there is an $\varepsilon > 0$ such that, for every analytic function $g(x')$, the integral

$$I(x_n) = \int u(x',x_n)g(x')\,dx'$$

is an analytic function when $|x_n| < \varepsilon$. Since by assumption, $I(x_n) = 0$ when $x_n \leqslant 0$ it will follow that $I(x_n) = 0$ when $|x_n| < \varepsilon$ and because the analytic functions are dense in the space of continuous functions we must have $u(x) = 0$ when $|x_n| < \varepsilon$.

In view of Lemma 1' we have to prove that if the functions ψ_N have support close to 0 and satisfy (6) then

$$|\widehat{\psi_N I}(\xi_n)| \leqslant C^{N+1} N^N (1 + |\xi_n|)^{-N}. \tag{9}$$

Since there is a conic neighborhood of $x = 0$, $\xi' = 0$, $\xi_n \neq 0$ which is disjoint from $WF_A(u)$ and since $g(x')$ is analytic it follows that, if the support of ψ_N is small enough, we can write $(\psi_N I)(x_n)$ as a finite sum of terms of the form

$$I_N(x_n) = \int u_N(x',x_n)g_N(x')\,dx',$$

where

$$|\hat{u}_N(\xi',\xi_n)| \leqslant C^{N+1} N^N (1 + |\xi'| + |\xi_n|)^{-N}, \tag{10}$$

when $|\xi'| \leqslant \delta|\xi_n|$ for some $\delta > 0$, and

$$|\hat{g}_N(\xi')| \leq C^{N+1} N^N (1 + |\xi'|)^{-N}. \tag{11}$$

Here $\hat{g}_N$ denotes the (n-1)-dimensional Fourier transform of g_N. In view of Parseval's formula we have

$$\hat{I}_N(\xi_n) = \int \hat{u}_N(\xi',\xi_n)\hat{g}_N(-\xi')d\xi'. \tag{12}$$

From (11) it follows that

$$|\hat{g}_N(\xi')| \leq C_1^{N+1} N^N (1 + |\xi'| + \delta|\xi_n|)^{-N},$$

when $|\xi'| \geq \delta|\xi_n|$. This together with (10) give

$$|\hat{u}_N(\xi',\xi_n)\hat{g}_N(-\xi')| \leq C_2^{N+1} N^N (1 + |\xi'| + |\xi_n|)^{-N}.$$

Therefore (9), with a suitable constant C, follows from (12).

In view of Theorem 3, the following classical uniqueness theorem is a consequence of Theorem 1'.

THEOREM 4. *Suppose that* Ω_o *is an open subset of* Ω *with* C^1*-boundary* $\partial\Omega_o$. *Denote by* N_o *a normal of* $\partial\Omega_o$ *at* $x_o \in \Omega$ *and let* $P(x,D)$ *be a linear differential operator with analytic coefficients such that* $P_m(x_o,N_o) \neq 0$. *Then there is a neighborhood* Ω' *of* x_o *such that if* $u \in D'(\Omega)$, $P(x,D)u = 0$ *in* Ω *and* $u = 0$ *in* Ω_o *then* $u = 0$ *in* Ω'.

By combining Theorem 3 with more precise results about "propagation of analyticity", corresponding to Theorem 2, it is possible to improve Theorem 4 considerably to obtain uniqueness also cases where the surface $\partial\Omega_o$ is no longer non-characteristic, i.e. we may have $P_m(x_o,N_o) = 0$.

SOME REFERENCES

[1] ANDERSSON, K.G., *Propagation of analyticity for solutions of differential equations of principal type*, Bull. Amer. Math. Soc., Vol. 78, 1972.

[2] HÖRMANDER, L., *Linear Differential Operators*, Actes Congres Intern. Math., Nice, 1970.

[3] HÖRMANDER, L., *Uniqueness theorems and wave front sets for solutions of linear differential equations with analytic coefficients*, Com. Pure Appl. Math., Vol. 24, 1971.

[4] HÖRMANDER, L., *On the existence and regularity of solutions of linear pseudo-differential equations*, L'Enseignement Math., Vol. 17, 1971.

[5] KAWAI, T., *Construction of local elementary solutions for linear partial differential operators with real analytic coefficients I - The case with real principal symbols*, Publ. R.I.M.S. Kyoto, Vol. 7, 1971.

[6] SATO, M., *Regularity of hyperfunction solutions of partial differential equations*, Actes Congres Intern. Math. Nice, 1970.

Lund University
Departement of Mathematics
Lund
SWEDEN

FUNDAMENTAL SOLUTIONS OF HYPOELLIPTIC BOUNDARY VALUE PROBLEMS

by

J. Barros Neto [1]

Let $P = P(D)$ be a hypoelliptic partial differential operator with constant coefficients in $\mathbb{R}^N$. Let $N = n+1$ and suppose, eventually after a change of variables, that the operator P can be written as follows

(1) $$P = P(D,D_t) = D_t^\sigma + a_1(D)D^{\sigma-1} + \dots + a_\sigma(D)$$

where $a_j(D)$ is, for every $j = 1,\dots,\sigma$, a partial differential operator with constant coefficients with respect to

$$D = (D_1,\dots,D_n), \quad D_j = \frac{1}{i}\frac{\partial}{\partial x_j}, \quad 1 \leqslant j \leqslant n, \quad \text{and} \quad D_t = \frac{1}{i}\frac{\partial}{\partial_t} .$$

Suppose further that the hypoelliptic partial differential operator is of *type* μ ($\mu \leqslant \sigma$), that is, the equation in τ

(2) $$P(\xi,\tau) = 0$$

has precisely μ *roots (counting multiplicities) with positive imaginary part, none of which is real,* for all $\xi \in \mathbb{R}^n$ with $|\xi|$ sufficiently large.

Let Ω be an open subset of $\mathbb{R}^{n+1}$ consisting of all vectors

$$(x,t) = (x_1,\dots,x_n,t) \in \mathbb{R}^{n+1}$$

with $t > 0$ and suppose that its boundary contains a plane piece ω contained in

---

[1] This article was written while the author was visiting the Universidade Federal de São Carlos (Brasil).

$$\mathbb{R}^n_o = \{(x,t) \in \mathbb{R}^{n+1} : t = 0\}.$$

DEFINITION. Let $P(D,D_t)$ be a hypoelliptic partial differential operator of type μ and let be given μ partial differential operators $Q_1(D,D_t),\dots,Q_\mu(D,D_t)$ with constant coefficients. We say that

$$(P(D,D_t)\ ;\ Q_1(D,D_t),\dots,Q_\mu(D,D_t))$$

defines a hypoelliptic boundary value problem in $\Omega \cup \omega$ if every function $u \in C^k(\Omega \cup \omega)$ (where k denotes the maximum order of the operators $P,Q_1,\dots,Q_\mu$) which is a solution of the boundary problem

$$(3) \qquad \begin{cases} P(D,D_t)u = f & \text{in } \Omega \\ Q_\nu(D,D_t)u_{|\omega} = g_\nu\ , & 1 \leqslant \nu \leqslant \mu \end{cases}$$

belongs to $C^\infty(\Omega \cup \omega)$.

In our papers [2] and [3] we proved that a necessary and sufficient condition for a boundary value problem to be hypoelliptic, in the sense above defined, is the existence of a parametrix with suitable regularity properties. By refining the arguments used in [2] and [3] we are going to show how it is possible to construct *fundamental solutions* of hypoelliptic boundary value problems and to characterize such problems by means of regularity properties of fundamental solutions.

THEOREM. $(P(D,D_t)\ ;\ Q_1(D,D_t),\dots,Q\ (D,D_t))$ *defines a hypoelliptic boundary value problem in* $\Omega \cup \omega$ *if and only if there are distributions* $K(x,t),K_1(x,t),\dots,K_\mu(x,t)$ *belonging to* $\mathcal{D}'(\mathbb{R}^{n+1}_+)$ *such that*

i) $K(x,t),K_1(x,t),\dots,K_\mu(x,t)$ *belong to* $C^\infty_c(\overline{\mathbb{R}^{n+1}_+} \setminus \{0\})$;

ii) $K(x,t)$ *is a solution of the boundary problem*

$$(4)\qquad \begin{cases} P(D,D_t)\ K(x,t) = \delta_x \times \delta_t \\ Q_\nu(D,D_t)\ K(x,t)\Big|_{\mathbb{R}^n_0} = 0, \quad 1 \leqslant \nu \leqslant \mu; \end{cases}$$

iii) *for every* $\nu = 1,\ldots,\mu$, $K_\nu(x,t)$ *is a solution of the boundary problem*

$$(5)\qquad \begin{cases} P(D,D_t)\ K_\nu(x,t) = 0 \\ Q_\ell(D,D_t)\ K_\nu(x,t)\Big|_{\mathbb{R}^n_0} = \delta_{\ell,\nu}\delta_x\ , \quad 1 \leqslant \ell \leqslant \nu. \end{cases}$$

PROOF. 1. Denote by $\tau_1,\ldots,\tau_\sigma$ all the roots of equation (2) and consider the $\mu \times \sigma$ matrix

$$(6)\qquad \begin{pmatrix} Q_1(\xi,\tau_1) & \cdots & Q_1(\xi,\tau_\sigma) \\ \cdots\cdots & \cdots & \cdots\cdots \\ Q_\mu(\xi,\tau_1) & \cdots & Q_\mu(\xi,\tau_\sigma) \end{pmatrix}$$

Let J denote any μ-tuple $(j_1,\ldots,j_\mu)$ of integers such that

$$1 \leqslant j_1 < \ldots \leqslant j_\mu \leqslant \sigma,$$

set

$$k^J_\xi(\tau) = \prod_{k=1}^{\mu}(\tau - \tau_{j_k}(\xi))$$

and consider the polynomial in $\xi = (\xi_1,\ldots,\xi_n)$ and $\tau = (\tau_1,\ldots,\tau_\sigma)$ defined by

$$C^J(\xi,\tau) = R(k_\xi^J;\ Q_1,\ldots,Q_\mu) = \frac{\det(Q_\nu(\xi,\tau_{j_k}(\xi)))_{\substack{1\leq\nu\leq\mu \\ 1\leq k\leq\mu}}}{\prod_{k<\ell}(\tau_{j_k}(\xi) - \tau_{j_\ell}(\xi))} .$$

Next, define

$$F(\xi,\tau) = \sum_j |C^J(\xi,\tau)|^2.$$

It is a polynomial in ξ and τ which, as easily seen, is symmetric in $\tau_1,\ldots,\tau_\sigma$. Therefore, by a theorem of elementary algebra, $F(\xi,\tau)$ is a polynomial in the coefficients of P. Since these coefficients are polynomials in ξ, it follows that $F(\xi,\tau)$ is also a polynomial in ξ which we shall denote by $\Phi(\xi)$.

2. The polynomial $\Phi(\xi)$ is *not identically zero*. Indeed, we recall that, by assumption, $P(\xi,\tau)$ is a hypoelliptic polynomial and that outside a compact subset of $\mathbb{R}^n$, equation (2) has μ roots (counting multiplicities) with positive imaginary part, none of which is real. Let A be the subset of $\mathbf{C}^n$ consisting of all $\zeta = (\zeta_1,\ldots,\zeta_n)$ for which the equation $P(\zeta,\tau)$ has μ roots (counting multiplicities) with positive imaginary part. It is clear that A is a non-empty open subset of $\mathbf{C}^n$. For every $\zeta\in A$ denote, for simplicity of notation, by $\tau_1(\zeta),\ldots,\tau_\mu(\zeta)$ the roots of $P(\xi,\tau) = 0$ with positive imaginary part, set

$$k_\zeta(\tau) = \prod_{j=1}^{\mu}(\tau - \tau_j(\zeta))$$

and define, as above,

$$(7') \qquad C(\zeta) = R(k_\zeta;\ Q_1,\dots,Q_\mu) = \frac{\det(Q_\nu(\zeta,\tau_j(\zeta)))_{\substack{1\leqslant\nu\leqslant\mu \\ 1\leqslant j\leqslant\mu}}}{\prod_{k<j}(\tau_j(\zeta)-\tau_k(\zeta))}\,.$$

It can be proved that $C(\zeta)$, called the *characteristic function of* the boundary problem, is analytic in A ([4]). Let

$$N = \{\zeta \in A : C(\zeta) = 0\}.$$

In [4], Hörmander has shown that the boundary problem $(P;\ Q_1,\dots,Q_\mu)$ is hypoelliptic in $\Omega \cup \omega$ if and only if the following condition holds

$$\text{(H)} \qquad \zeta \in N, \quad |\zeta| \longrightarrow +\infty \quad \text{implies} \quad |\mathrm{Im}\ \zeta| \longrightarrow +\infty.$$

As a consequence of (H), we can see that there is a constant $M > 0$ such that for all $\xi \in \mathbb{R}^n$ with $|\xi| > M$, we have $C(\xi) \neq 0$. Therefore, $\Phi(\xi)$ is not identically zero.

Define, for every $\nu = 1,\dots,\mu$,

$$(8) \qquad H_\nu(\xi,t) = \frac{R(k_\xi;Q_1(\xi,\tau(\xi)),\dots,e^{it\tau(\xi)},\dots,Q_\mu(\xi,\tau(\xi)))}{C(\xi)}$$

for all $|\xi| > M$ and $t \geqslant 0$, where

$$R(k_\xi;Q_1(\xi,\tau(\xi),\dots,e^{it\tau(\xi)},\dots,Q_\mu(\xi,\tau(\xi)))$$

indicates that in the determinant appearing in (7') the νth row has been replaced by

$$(e^{it\tau_1(\xi)},\dots,e^{it\tau_\mu(\xi)}).$$

It is easy to see that every $H_\nu(\xi,t)$, $1 \leqslant \nu \leqslant \mu$, is a solution of the initial value problem

$$\begin{cases} P(\xi,D_t)\, H_\nu(\xi,t) = 0 \\ Q_\ell(\xi,D_t)\, H_\nu(\xi,t)\Big|_{t=0} = \sigma_{\ell,\nu}, \quad 1 \leqslant \ell \leqslant \mu. \end{cases}$$

3. However, $\Phi(\xi)$ may vanish in the ball

$$\bar{B}(0,M) = \{\xi \in \mathbb{R}^n : |\xi| \leqslant M\}.$$

Proceeding as in Hörmander [5], there is a finite set A' of vectors $\theta \in \mathbb{R}^n$, $|\theta| \leqslant \varepsilon$, and a finite open convering $(V_\theta)_{\theta\in A'}$, of $\bar{B}(0,M)$, such that

$$F(\xi + z\theta) \neq 0, \quad \forall\ |z| = 1, \quad \forall\ \xi \in V_\theta.$$

This easily implies the existence of a finite subset $A \subset A'$ and a finite open covering $(U_\theta)_{\theta\in A}$ of $\bar{B}(0,M)$ such that, to every $\theta \in A$ it corresponds a finite covering of the circle $\{z \in \mathbf{C} : |z| = 1\}$ by open arcs $(\Gamma_{\theta,j})_{1\leqslant j\leqslant L_\theta}$ and a finite set of μ-tuples $(J_j)_{1\leqslant j\leqslant L_\theta}$, such that

$$C^{J_j}(\xi + z\theta\ ,\ \tau(\xi + z\theta)) \neq 0,\ \forall\ \xi \in U_\theta\ ,\ \ \forall\ z \in \Gamma_{\theta,j}\ .$$

Define, for every $\nu = 1,\ldots,\mu$,

(8')
$$H_\nu^{J_j}(\xi + z\theta\ ,\ t) =$$
$$\frac{R(k^{J_j}_{\xi+z\theta};Q_1(\xi+z\theta,\tau(\xi+z\theta)),\ldots,e^{it\tau(\xi+z\theta)},\ldots,Q_\mu(\xi+z\theta,\tau(\xi+z\theta)))}{C^{J_j}(\xi+z\theta\ ,\ \tau(\xi+z\theta))}$$

$\xi \in U_\theta$, $z \in \Gamma_{\theta,j}$. It is clear that every $H_\nu^{J_j}(\xi + z\theta, t)$, $1 \leqslant \nu \leqslant \mu$, is a solution of the value problem

$$(9') \quad \begin{cases} P(\xi + z\theta, D_t)\, H_\nu^{J_j}(\xi + z\theta, t) = 0 \\ Q_\ell(\xi+z\theta, D_t) H_\nu^{J_j}(\xi+z\theta, t')\big|_{t=0} = \delta_{\ell,\nu}, \quad 1 \leqslant \ell \leqslant \mu. \end{cases}$$

4. For every θ, let $(\Psi_{\theta,j})_{1\leqslant j\leqslant L_\theta}$ be a partition of unity subordinated to the covering $(\Gamma_{\theta,j})_{1\leqslant j\leqslant L_\theta}$ and let $(\phi_\theta)_{\theta\in A}$ be a partition of unity subordinated to the covering $(U_\theta)_{\theta\in A}$. Define, for every $\nu = 1,\dots,\mu$

$$(10) \quad \langle K_\nu(x,t), \check{f}(x)\rangle = (2\pi)^{-n} \sum_{\theta \in A} \int_{|\xi|\leqslant M} \phi_\theta(\xi) \left\{ \frac{1}{2\pi i} \int_{|z|=1} \sum_{J=1}^{L_\theta} \Psi_{\theta,j}(z) \frac{H_\nu^{J_j}(\xi+z\theta,t)\hat{f}(\xi+z\theta)\,dz}{z} \right\} d\xi$$

$$+ (2\pi)^{-n} \int_{|\xi|>M} H_\nu(\xi,t)\hat{f}(\xi)\,d\xi,$$

where $f \in C_c^\infty(\mathbb{R}^n)$ and $\check{f}(x) = f(-x)$. The integrals defining $K_\nu(x,t)$ do converge. We recall that the last integral has been already studied in our papers [2] and [3] where it has been shown that it is an absolutely convergent one. It is easy to see that $K_\nu(x,t)$ satisfies (5).

The existence of $K(x,t)$ satisfying (4) is easy to prove. It suffices to take a fundamental solution of the hypoelliptic operator $P(D, D_t)$ and to modify it in a suitable way in order that the boundary conditions (4) be satisfied.

6. To prove that $K(x,t)$, $(K_\nu(x,t))_{1\leqslant\nu\leqslant\mu}$ satisfy the required property i), we refer to our paper [2]. Also, in the proof that the existence of distributions K, $(K_\nu)_{1\leqslant\nu\leqslant\mu}$ satisfying conditions i), ii) and iii) implies that the boundary value problem $(P;Q_1,\ldots,Q_\mu)$ is hypoelliptic is analogous to the proof concerning regularity of parametrices of the boundary value problem ([2]), q.e.d.

As a final remark, let us mention that it is possible that the distribution K, $(K_\nu)_{1\leqslant\nu\leqslant\mu}$ belong to suitable Gevrey classes in $\overline{\mathbf{R}^{n+1}_+}\setminus\{0\}$ ([1], [3]).

Rutgers University
Department of Mathematics
New Brunswick, N.J. 08903
U.S.A.

BIBLIOGRAPHY

[1] R.ARTINO, "Gevrey classes and hypoelliptic boundary value problem", Thesis, Rutgers University.

[2] J.BARROS NETO, *The parametrix of a regular hypoelliptic boundary value problems*, Ann. della Sc. Normale Superiori di Pisa, Serie III, vol. XXVI, Fasc. 1, (1972), 247-268.

[3] J.BARROS NETO, *On regular hypoelliptic boundary problems*, Journ. Math. Anal. and Appl., vol.**41**, Nº2, 1973, pp.508-530.

[4] L.HÖRMANDER, *On the regularity of the solutions of boundary problems*, Acta Math. **99** (1958), 225-264.

[5] L.HÖRMANDER, Linear Partial differential operators, Springer Verlag, Berlin, 1963.

THE GREEN FUNCTION OF A LINEAR DIFFERENTIAL EQUATION WITH LATERAL CONDITION

by

Chaim Samuel Hönig

Let E be a Banach space. We consider systems of the form

$$\text{(L)} \qquad L[y] \equiv y' + Ay = f$$

$$\text{(F)} \qquad F[y] = c$$

where $y \in C^{(1)}([a,b],E)$, $f \in C([a,b],E)$, $A \in C([a,b],L(E))$, $F \in L[C([a,b],E),E]$ and $c \in E$. When the system has one and only one solution for any $f \in C([a,b],E)$ and $c \in E$, we show that it has a Green function, that is, a function $G: [a,b] \times [a,b] \longrightarrow L(E,E'')$ such that $y \in C^{(1)}([a,b],E)$ is the solution of $L[y] = f$ and $F[y] = 0$ if and only if

$$y(t) = \int_a^b G(t,s) f(s) ds.$$

We exhibit the relations between G, A and F.

By the usual transformations a linear differential equation of order n, or a system of such equations, can be reduced to the form (L). (F) is called a *lateral condition* or a *generalized boundary condition*; initial conditions and boundary conditions are particular instances of lateral conditions. In §3 we give extensions of our results to more general situations.

§ 1 - ANALYTIC PRELIMINARIES

We consider always vector spaces over the complex field **C** but all results are valid for real vector spaces.

1. Given an interval $[a,b]$ of the real line, a *division* of $[a,b]$

is a finite sequence $d\colon t_o = a < t_1 < \dots < t_n = b$. We write

$$|d| = n \quad \text{and} \quad \Delta d = \sup\{|t_i - t_{i-1}| \mid i = 1,2,\dots,|d|\};$$

D denotes the set of all divisions of $[a,b]$.

2. Let E, F and G be vector spaces, F and G normed; let

$$B\colon (x,y) \in E \times F \longmapsto x \cdot y = B(x,y) \in G$$

be a bilinear mapping and $\alpha\colon [a,b] \longrightarrow E$. For $d \in D$ we define

$$SB_d[\alpha] = \sup\{\| \sum_{i=1}^{|d|} |\alpha(t_i) - \alpha(t_{i-1})| \cdot y_i \| \mid y_i \in F, \|y_i\| \leqslant 1\}$$

and

$$SB[\alpha] = \sup\{SB_d[\alpha] \mid d \in D\}.$$

We say that α is of *bounded* B-*variation* if $SB[\alpha] < \infty$ and we write $\alpha \in SB([a,b],E)$.

For $\alpha\colon [a,b] \longrightarrow E$ and $f\colon [a,b] \longrightarrow F$ we define

$$\int_a^b d\alpha(t) \cdot f(t) = \lim_{\Delta d \to 0} \sum_{i=1}^{|d|} [\alpha(t_i) - \alpha(t_{i-1})] \cdot f(\xi_i) \in G,$$

where $\xi_i \in [t_{i-1}, t_i]$, if the limit exists. When $E = \mathbf{C}$, $F = G$, $B(\lambda,x) = \lambda x$ and $\alpha(t) = t$ we get the Riemann integral of vector valued functions, $\int_a^b f(t)dt$.

THEOREM 1.1. *Let* G *be a Banach space; for*

$$\alpha \in SB([a,b],E) \quad \textit{and} \quad f \in C([a,b],F)$$

there exists

$$F_\alpha[f] = \int_a^b d\alpha(t)\cdot f(t)$$

and we have $\|F_\alpha[f]\| \leq SB[\alpha]\,\|f\|$, *that is*

$$F_\alpha \in L\big[C([a,b],F),G\big] \quad \textit{and} \quad \|F_\alpha\| \leq SB[\alpha].$$

PROPOSITION 1.2. *Let* E, F *and* G *be normed spaces,* G *complete; let* $B\colon E\times F \longrightarrow G$ *be a continuous bilinear mapping. For*

$$\alpha \in SB([a,b],E) \quad \textit{and} \quad f \in C^{(1)}([a,b],F)$$

there exists

$$\int_a^b \alpha(t)\cdot f'(t)\,dt = \int_a^b \alpha(t)\cdot df(t).$$

PROOF. From theorem 1.1 it follows, integrating by parts, that the second integral exists. To prove the existence of the first integral and the equality it is enough to show that the sums

$$\sum_{i=1}^{|d|} \alpha(\xi_i)\cdot f'(\xi_i)(t_i - t_{i-1})$$

are arbitrary close to the corresponding sums of the second integral. We have

$$\Big\|\sum_{i=1}^{|d|} \alpha(\xi_i)\cdot[f(t_i)-f(t_{i-1})] - \sum_{i=1}^{|d|} \alpha(\xi_i)\cdot f'(\xi_i)(t_i-t_{i-1})\Big\| =$$

$$= \Big\|\sum_{i=1}^{|d|} \alpha(\xi_i)\cdot[f(t_i) - f(t_{i-1}) - f'(\xi_i)(t_i - t_{i-1})]\Big\| \leq$$

$$\leq \|B\|\,\|\alpha\| \sum_{i=1}^{|d|} \omega_i(f')(t_i - t_{i-1}),$$

where

$$\omega_i(f') = \sup\{\|f'(\eta) - f'(\xi)\| \mid \eta,\xi \in [t_{i-1},t_i]\},$$

(see [F], (8.6.2)). Using the uniform continuity of f' on $[a,b]$ the result follows.

OBSERVATION. Vector valued integrals of the form

$$\int_a^b d\alpha(t)\cdot f(t) \quad \text{or} \quad \int_a^b g(t)dt$$

must be handled carefully; they are defined as Riemann integrals and don't satisfy the Darboux condition nor do they exist in the sense of Bochner-Lebesgue, that is, they are not obtained as continuous extensions from integrals of continuous functions. Hence their properties must be proved directly and many of them are lost. For instance, the existence of $\int_a^b f(t)dt$ does not imply the existence of $\int_a^b \|f(t)\|dt$ and f may even be non-measurable. Neither does the existence of $\int_a^b f(t)dt$ and $\int_a^b g(t)dt$, where $f\colon [a,b] \longrightarrow E$ and $g\colon [a,b] \longrightarrow F$, imply the existence of $\int_a^b f(t)\cdot g(t)dt \in G$.

3. EXAMPLE A: $BV([a,b],X)$ - Let X be a Banach space; we take $E = X$, $F = X'$ its dual, $G = \mathbf{C}$ and $B(x,x') = \langle x,x'\rangle$. For

$$\alpha\colon [a,b] \longrightarrow X$$

and $d \in D$ we define

$$V_d[\alpha] = \sum_{i=1}^{|d|} \|\alpha(t_i) - \alpha(t_{i-1})\|;$$

from the Hahn-Banach theorem it follows that

$$V_d[\alpha] = \sup\{|\sum_{i=1}^{|d|} \langle\alpha(t_i)-\alpha(t_{i-1}),x_i'\rangle| \mid x_i' \in X', \|x_i'\| \leq 1\}$$

and hence $V[\alpha] = \sup\{V_d[\alpha] \mid d \in D\}$ is the usual (strong) *variation* of α and the functions of bounded B-variation are, in this case, the functions of (strong) bounded variation $(V[\alpha] < \infty)$. We write $\alpha \in BV([a,b],X)$ if $V[\alpha] < \infty$.

By $\widetilde{BV}_o([a,b],X)$ we denote the space of all functions

$$\alpha \in BV([a,b],X)$$

such that $\alpha(a) = 0$ and $\alpha(t+) = \alpha(t)$ for $t \in]a,b[$. Endowed with the norm $V[\alpha]$, $\widetilde{BV}_o([a,b],X)$ is a Banach space. We write

$$\widetilde{BV}_o([a,b]) = \widetilde{BV}_o([a,b],\mathbf{C}).$$

In the usual way one proves the following

THEOREM 1.3 (Riesz). $C([a,b],X)' \simeq \widetilde{BV}_o([a,b],X')$; *that is, the mapping*

$$\alpha \in \widetilde{BV}_o([a,b],X') \longmapsto F_\alpha \in C([a,b],X)'$$

is a linear isometry (i.e., $\|F_\alpha\| = V[\alpha]$*) of the first Banach space onto the second.*

(We recall that according to Theorem 1.1, for $f \in C([a,b],X)$ we have

$$F_\alpha[f] = \langle f,F_\alpha\rangle = \int_a^b \langle f(t),d\alpha(t)\rangle = \lim_{\Delta d \to 0} \sum_{i=1}^{|d|} \langle f(\xi_i),\alpha(t_i)-\alpha(t_{i-1})\rangle$$

where $\xi_i \in [t_{i-1},t_i]$.)

EXAMPLE B: $BW([a,b],X)$. Let X be a Banach space; we take $E = X$, $F = \mathbf{C}$, $G = X$ and $B(x,\lambda) = \lambda x$. For $\alpha: [a,b] \longrightarrow X$ and $d \in D$ we define

$$W_d[\alpha] = \sup\{|\sum_{i=1}^{|d|} \lambda_i[\alpha(t_i)-\alpha(t_{i-1})]| \mid \lambda_i \in \mathbf{C}, |\lambda_i| \leq 1\}$$

and $W[\alpha] = \sup\{W_d[\alpha] \mid d \in D\}$. We say that α is of *weak bounded variation*, and write $\alpha \in BW([a,b],X)$, if $W[\alpha] < \infty$. This definition is equivalent to the usual one because from the principle of uniform boundedness it follows that

$$BW([a,b],X) = \{\alpha \in X^{[a,b]} \mid x' \circ \alpha \in BV([a,b]) \text{ for all } x' \in X'\}$$

and

$$W[\alpha] = \sup\{V[x' \circ \alpha] \mid x' \in X', \|x'\| \leq 1\}$$

(See for instance [H-P], Theorem 3.2.2.)

If X is a Banach space by $\widetilde{BW}_o([a,b],X')$ we denote the set of all functions $\alpha \in BW([a,b],X')$ such that $x \circ \alpha \in \widetilde{BV}_o([a,b])$ for all $x \in X$; in other words, $\alpha \in BW([a,b],X')$, $\alpha(a) = 0$ and the function

$$t \in]a,b[\longmapsto \alpha(t) \in X'_{\sigma(X',X)}$$

is continuous on the right. Endowed with the norm $W[\alpha]$, $\widetilde{BW}_o([a,b],X')$ is a Banach space.

THEOREM 1.4. $L[C([a,b]),X'] \simeq \widetilde{BW}_o([a,b],X')$; *that is, the mapping*

$$\alpha \in \widetilde{BW}_o([a,b],X') \longmapsto F_\alpha \in L[C([a,b]),X']$$

is a linear isometry (i.e., $\|F_\alpha\| = W[\alpha]$*) of the first Banach space onto the second.*

(We recall that, according to Theorem 1.1 for $\phi \in C([a,b])$ we have

$$F_\alpha[\phi] = \int_a^b \phi(t)\,d\alpha(t).)$$

In order to construct the Green function of the system (L), (F) we need a representation of $F \in L[C([a,b],E),E]$ by an integral generalizing the representations given by Theorems 1.3 and 1.4. To this end we introduce the following

EXAMPLE C: $SV([a,b],L(X,Y))$. Let X and Y be Banach spaces; we take $E = L(X,Y)$, $F = X$, $G = Y$ and $B(u,x) = u(x)$, where $u \in L(X,Y)$ and $x \in X$. For $\alpha\colon [a,b] \longrightarrow L(X,Y)$ and $d \in D$ we define

$$SV_d[\alpha] = \sup\{\|\sum_{i=1}^{|d|} [\alpha(t_i)-\alpha(t_{i-1})]x_i\| \mid x_i \in X,\ \|x_i\| \leqslant 1\}$$

and

$$SV[\alpha] = \sup\{SV_d[\alpha] \mid d \in D\};$$

we say that α is of *bounded semi-variation* if $SV[\alpha] < \infty$ and write $\alpha \in SV([a,b],L(X,Y))$. (See for instance [D].)

We have $BV([a,b],L(X,Y)) \subset SV([a,b],L(X,Y)) \subset BW([a,b],L(X,Y))$.

When $X = \mathbf{C}$ we have

$$SV([a,b],L(\mathbf{C},Y)) = BW([a,b],Y);$$

when $Y = \mathbf{C}$ we have

$$SV([a,b],L(X,\mathbf{C})) = BV([a,b],X').$$

Let X and Z be Banach spaces and $\alpha: [a,b] \longrightarrow L(X,Z')$; for $x \in X$ and $z \in Z$ we define

$$(z\circ\alpha)(x): [a,b] \longrightarrow \mathbf{C}$$

by

$$[(z\circ\alpha)(x)](t) = \langle z,\alpha(t)x\rangle.$$

We denote by $\widetilde{SV}_o([a,b],L(X,Z'))$ the set of all functions

$$\alpha \in SV([a,b],L(X,Z'))$$

such that $(z\circ\alpha)(x) \in \widetilde{BV}_o([a,b])$ for every $x \in X$ and $z \in Z$; that is, $\alpha \in SV([a,b],L(X,Z'))$, $\alpha(a) = 0$ and for every $x \in X$ the function

$$t \in]a,b[\longmapsto \alpha(t)x \in Z'_{\sigma(Z',Z)}$$

is continuous on the right. Endowed with the norm $SV[\alpha]$, $\widetilde{SV}_o([a,b],L(X,Z'))$ is a Banach space. One has (see for instance [B-K], Satz 11):

THEOREM 1.5. *Let* X *and* Z *be Banach spaces; the mapping*

$$\alpha \in \widetilde{SV}_o([a,b],L(X,Z')) \longmapsto F_\alpha \in L[C([a,b],X),Z']$$

is a linear isometry (that is, $\|F_\alpha\| = SV[\alpha]$*) of the first Banach space onto the second.*

(Recall that, according to Theorem 1.1, for $f \in C([a,b],X)$ we have

$$F_\alpha[f] = \int_a^b d\alpha(t)f(t).)$$

COROLLARY. *Let* X *and* Y *be Banach spaces. For every*

$$F \in L[C([a,b],X),Y]$$

there exists one and only one $\alpha \in \widetilde{SV}_o([a,b],L(X,Y''))$ *such that* $F=F_\alpha$. *We write* $\alpha_F = \alpha$.

§ 2 - THE GREEN FUNCTION

1. Given the differential operator L defined in the introduction, for every $s \in [a,b]$ the function $R \in C^{(1)}([a,b],L(E))$ solution of

$$\frac{dR}{dt} + A \circ R = 0$$

such that $R(s) = I_E$ (identical automorphism of E) is denoted by R_s and is called the *resolvent* of L. We write $R(t,s) = R_s(t)$, where $t \in [a,b]$.

For the following three propositions see, for instance, [B], Chap. IV, §2 or [C].

PROPOSITION 2.1. *The function* $y_{s,x} \in C^{(1)}([a,b],E)$ *solution of* $L[y]=0$ *such that* $y(s) = x$, *where* $x \in E$, *is given by* $y_{s,x}(t) = R_s(t)x$.

PROPOSITION 2.2. *For* $t,s,\sigma \in [a,b]$ *we have* $R(t,s) \circ R(s,\sigma) = R(t,\sigma)$, $R(s,t) = R(t,s)^{-1}$ *and* $R(s,s) = I_E$.

PROPOSITION 2.3. *For* $f \in C([a,b],E)$ *and* $c \in E$ *the solution of* $L[y]=f$, $y(s) = c$ *is given by*

$$y(t) = R(t,s)c + \int_s^t R(t,\sigma)f(\sigma)d\sigma.$$

2. Given $F \in L[C([a,b],E),E]$ and $s \in [a,b]$, for every $x \in E$ we define $F[R_s]x = F[R_s x]$, hence $F[R_s] \in L(E)$. It is easy to show that

$$F[R_s] = \int_a^b d\alpha(t) \circ R(t,s)$$

where $\alpha = \alpha_F$ (see Corollary of Theorem 1.5). We write

$$J_s = J(s) = F[R_s] = F_t[R(t,s)].$$

THEOREM 2.4. *The following properties are equivalent:*

1) *For every* $f \in C([a,b],E)$ *and* $c \in E$ *the system* $L[y] = f$, $F[y] = c$ *has one and only one solution* $y \in C^{(1)}([a,b],E)$;
2) *For every* $c \in E$ *the system* $L[y] = 0$, $F[y] = c$ *has one and only one solution* $y \in C^{(1)}([a,b],E)$;
3) *The mapping*

$$y \in \{u \in C^{(1)}([a,b],E) \mid L[u] = 0\} \longmapsto F[y] \in E$$

is an isomorphism of the first Banach space onto the second;

4) *For every* $s \in [a,b]$ *we have* $J_s = F[R_s] \in \mathrm{Aut}(E)$;
5) *There is an* $s \in [a,b]$ *such that* $J_s \in \mathrm{Aut}(E)$.

PROOF. We have obviously 1) $\Longrightarrow$ 2). 2) $\Longrightarrow$ 1): by proposition 2.3 there exists a solution $\tilde{y} \in C^{(1)}([a,b],E)$ of $L[\tilde{y}] = f$ and by 2) there exists one and only one solution $z \in C^{(1)}([a,b],E)$ of $L[z] = 0$, $F[z] = c - F[\tilde{y}]$. Then $y = \tilde{y} + z$ is the unique solution of $L[y] = f$, $F[y] = c$. 2) $\Longrightarrow$ 3) is immediate. 3) $\Longrightarrow$ 4): the mapping

$$x \in E \longmapsto y_{s,x} \in L^{-1}(0)$$

(see Proposition 2.1) is an isomorphism, and by 3) the mapping

$$y \in L^{-1}(0) \longmapsto F[y] \in E$$

is an isomorphism too; hence the same is true for the composed mapping

$$x \in E \longmapsto J_s x = F[R_s]x = F[R_s x] \in E.$$

4) $\Longrightarrow$ 5) is obvious. 5) $\Longrightarrow$ 3): if $J_s = F[R_s] \in \mathrm{Aut}(E)$ then the mapping

$$x \in E \longmapsto F[R_s]x = F[R_s x] = F[y_{s,x}] \in E$$

is an isomorphism; but it is composed of the isomorphism

$$x \in E \longmapsto y_{s,x} \in L^{-1}(0)$$

and the homomorphism $y \in L^{-1}(0) \longmapsto F[y] \in E$ and hence this last one is an isomorphism too.

From now on we suppose that the equivalent properties of Theorem 2.4 are verified.

PROPOSITION 2.5. *For* $t,s \in [a,b]$ *we have* $R(t,s) = J_t^{-1} \circ J_s$.

PROOF. It is enough to show that $J_t \circ R(t,s) = J_s$. We have

$$J_s = F_\tau[R(\tau,s)] = F_\tau[R(\tau,t) \circ R(t,s)]$$

and for $S \in L(E)$ we have

$$F_\tau[R(\tau,t) \circ S] = F_\tau[R(\tau,t)] \circ S$$

because for every $x \in E$ we have

$$F_\tau[R(\tau,t)] \circ Sx = F_\tau[R(\tau,t)Sx] = F_\tau[R(\tau,t) \circ S]x.$$

Hence

$$F_\tau[R(\tau,t) \circ R(t,s)] = F_\tau[R(\tau,t)] \circ R(t,s) = J_t \circ R(t,s).$$

PROPOSITION 2.6. $F_t[J(t)^{-1}] = I_E$.

PROOF. It suffices to remark that by Proposition 2.5 one has

$$J_s = F_t[R(t,s)] = F_t[J_t^{-1} \circ J_s] = F_t[J_t^{-1}] \circ J_s.$$

PROPOSITION 2.7. $\frac{d}{dt} J(t)^{-1} + A(t) \circ J(t)^{-1} = 0.$

PROOF. By Proposition 2.5 we have $J(t)^{-1} = R(t,s) \circ J(s)^{-1}$ and by the definition of the resolvent we have

$$\frac{dR(t,s)}{dt} + A(t) \circ R(t,s) = 0,$$

hence the result.

3. The main theorem

THEOREM 2.8. *If the properties of Theorem 2.4 are verified then*

$$y \in C^{(1)}([a,b],E)$$

is the solution of the system $L[y] = y' + Ay = f$, $F[y] = c$ *if and only if*

$$\text{(G)} \qquad y(t) = J(t)^{-1}c + \int_a^b G(t,s)f(s)ds$$

where

$$G(t,s) = \hat{J}(t)^{-1} \circ \left[\int_a^s d\alpha_F(\tau) \circ J(\tau)^{-1} - Y(s-t)I_E\right] \circ J(s),$$

$\hat{J}(t)^{-1} \in L(E'')$ *being the bitransposed of* $J(t)^{-1} \in L(E)$ *and* Y *the Heavside function* ($Y(\sigma) = 1$ if $\sigma \geq 0$ and $Y(\sigma) = 0$ if $\sigma < 0$). *Moreover we have:*

(i) $G(t,s) \in L(E,E'')$;

(ii) $G(s+,s) - G(s-,s) = I_E$ *for every* $s \in]a,b[$;

(iii) $G(t,b) = 0$; $G(a,a) = -I_E$ *and* $G(t,a) = 0$ *for* $a < t \leq b$;

(iv) *For every fixed* $s \in [a,b]$, G *is a continuous function of* t, *for* $t \neq s$;

(v) *For every fixed* $t \in [a,b]$ *and every* $x \in E$ *the function*

$$s \in]a,b[\longmapsto G(t,s)x \in E''_{\sigma(E'',E')}$$

is continuous on the right;

(v) *The function* G *with these properties is unique.*

PROOF. Let $y \in C^{(1)}([a,b]),E)$ be a solution of the system

$$L[y] = f, \quad F[y] = c;$$

by Proposition 2.3 we have

$$y(\tau) = R(\tau,t)y(t) + \int_t^\tau R(\tau,s)f(s)ds =$$

$$= R(\tau,t)y(t) + \int_a^\tau R(\tau,s)f(s)ds - \int_a^t R(\tau,s)f(s)ds.$$

Applying F we get

$$c = F_\tau[y(\tau)] = F_\tau[R(\tau,t)y(t)] +$$

$$+ F_\tau\left(\int_a^\tau R(\tau,s)f(s)\right) - \int_a^t F_\tau[R(\tau,s)f(s)]ds =$$

$$= J(t)y(t) + F_\tau\left[\int_a^\tau R(\tau,s)f(s)ds\right] - \int_a^t J(s)f(s)ds.$$

By the Corollary of Theorem 1.5 we have

$$F_\tau\left[\int_a^\tau R(\tau,s)f(s)ds\right] = \int_a^b d\alpha(\tau)\left[\int_a^\tau R(\tau,s)f(s)ds\right],$$

where $\alpha = \alpha_F$. To proceed with the proof we need the following

LEMMA 2.9. $\int_a^b d\alpha(\tau) \circ \left[\int_a^\tau R(\tau,s)f(s)ds\right] = \int_a^b \left[\int_s^b d\alpha(\tau) \circ R(\tau,s)\right] f(s)ds.$

PROOF. Integrating by parts and applying Propositions 1.2 and 2.2 we have

(*) $$\int_a^b d\alpha(\tau) \circ \left[\int_a^\tau R(\tau,s)f(s)ds\right] = \alpha(b) \circ \int_a^b R(b,s)f(s)ds -$$

$$- \int_a^b \alpha(\tau)f(\tau)d\tau + \int_a^b \alpha(\tau)\left[\int_a^\tau A(\tau)R(\tau,s)f(s)ds\right]d\tau$$

and

(**) $$\int_a^b \left[\int_s^b d\alpha(\tau) \circ R(\tau,s)\right] f(s)ds = \int_a^b \alpha(b) \circ R(b,s)f(s)ds -$$

$$- \int_a^b \alpha(s)f(s)ds + \int_a^b \left[\int_s^b \alpha(\tau) \circ A(\tau)R(\tau,s)d\tau\right] f(s)ds.$$

Comparing (*) and (**) we see that the Lemma is a consequence of the equality

$$\int_a^b \alpha(\tau)\left[\int_a^\tau A(\tau)R(\tau,s)f(s)ds\right]d\tau = \int_a^b\left[\int_s^b \alpha(\tau)A(\tau)R(\tau,s)d\tau\right]f(s)ds,$$

which, with the use of Proposition 2.5, follows from

$$\int_c^d \alpha(\tau)A(\tau)J(\tau)^{-1}\left[\int_e^f J(s)f(s)ds\right]d\tau = \int_e^f\left[\int_c^d \alpha(\tau)A(\tau)J(\tau)^{-1}d\tau\right]J(s)f(s)ds$$

where $[c,d] \times [e,f] \subset [a,b] \times [a,b]$.

Continuation of the proof of Theorem 2.8: By Lemma 2.9 we have

$$c = J(t)y(t) + \int_a^b\left[\int_s^b d\alpha(\tau) \circ R(\tau,s)\right]f(s)ds - \int_a^t J(s)f(s)ds$$

$$= J(t)y(t) - \int_a^b\left[\int_a^s d\alpha(\tau) \circ R(\tau,s)\right]f(s)ds + \int_t^b J(s)f(s)ds$$

since

$$J(s) = \int_a^b d\alpha(\tau) \circ R(\tau,s).$$

Hence, with the use of Proposition 2.5 we get

$$y(t) = J(t)^{-1}\{c + \int_a^b\left[\int_a^s d\alpha(\tau) \circ J(\tau)^{-1}\right]J(s)f(s)ds - \int_t^b J(s)f(s)ds\}$$

and finally

$$(***) \qquad y(t) = J(t)^{-1}c + \int_a^b G(t,s)f(s)ds$$

where

$$G(t,s) = \hat{J}(t)^{-1} \circ \left[\int_a^s d\alpha(\tau) \circ J(\tau)^{-1} - Y(s-t) I_E\right] \circ J(s).$$

From this expression for G follow immediately the properties (i) to (v); the uniqueness of G will follow from Theorem 2.10 below.

Conversely: given $f \in C([a,b],E)$ and $c \in E$, y defined by (***) is the solution of the system $L[y] = f$, $F[y] = c$ since by hypothesis this system has one and only one solution and we proved above that this solution is given by (***).

EXAMPLE. Let us take $E = C([a,b])$, $L[y] = y'$ and $F \in L[C([a,b],E),E]$ such that $F[f](t) = (f(t))(t)$ for $f \in C([a,b],E)$ and $t \in [a,b]$. In this case the Green function is such that for $\phi \in E$ we have

$$G(t,s)\phi = \begin{cases} -I_E & \text{if } s = a = t \\ 0 & \text{if } s = a < t \leqslant b \\ [\chi_{[a,s]} - I_E]\phi & \text{if } a < s \leqslant b \text{ and } a \leqslant t \leqslant s \\ \chi_{[a,s]}\phi & \text{if } a < s < t \leqslant b \end{cases}$$

where

$$\chi_{[a,s]}(\sigma) = \begin{cases} 1 & \text{if } a \leqslant \sigma \leqslant s \\ 0 & \text{if } s < \sigma \leqslant b. \end{cases}$$

Hence $G(t,s) \in L(E,E")$ but $G(t,s) \notin L(E)$.

4. $L_1([a,b],E)$ denotes the set of (equivalence classes of) measurable functions $f\colon [a,b] \longrightarrow E$ such that

$$\|f\|_1 = \int_a^b \|f(t)\|\,dt < \infty,$$

the integral taken in the sense of Lebesgue. $L_1([a,b],E)$ is the completion of $C([a,b],E)$ with respect to the norm $\|f\|_1$. $L_1^{(1)}([a,b],E)$ denotes the set of functions $g\colon [a,b] \longrightarrow E$ that are primitives of functions from $L_1([a,b],E)$ (and hence there exists $g' \in L_1([a,b],E)$). $L_1^{(1)}([a,b],E)$ is a Banach space when endowed with the norm

$$\|g\|_1^{(1)} = \|g\|_1 + \|g'\|_1.$$

The Green function in Theorem 2.8 is bounded (by (iv), (ii) and (v)) and by continuous extension, we get the

THEOREM 2.10. *With the notations and hypothesis of Theorem 2.8 we have: $y \in L_1^{(1)}([a,b],E)$ is the solution of the system* $L[y] = f$, $F[y] = c$, *where* $f \in L_1([a,b],E)$ *and* $c \in E$, *if and only if*

$$y(t) = J(t)^{-1}c + \int_a^b G(t,s)f(s)\,ds$$

(now the integral is taken in the sense of Bochner-Lebesgue, that is, it is defined by continuous extension from $C([a,b],E)$ to $L_1([a,b],E)$).

This theorem allows us to prove the uniqueness of the Green function satisfying (i) to (v) of Theorem 2.8. Indeed, if there were two such functions, G_1 and G_2, then taking $H = G_1 - G_2$, from the uniqueness of the solutions of the system $L[y] = f$, $F[y] = c$ it follows that

$$\int_a^b H(t,s)f(s)ds = 0$$

for all $f \in L_1([a,b],E)$. Let us prove that $H(t,s) = 0$: by (iii) this is true for $s = a$ and $s = b$; for $s \in]a,b[$ we define

$$\phi_n(\sigma) = \begin{cases} n & \text{if} \quad s \leqslant \sigma \leqslant s + \frac{1}{n} \\ 0 & \text{elsewhere} \end{cases}$$

and for $x \in E$ we put $f_n(\sigma) = \phi_n(\sigma)x$; then $f \in L_1([a,b],E)$ and

$$0 = \int_a^b H(t,\sigma)f_n(\sigma)d\sigma = n\int_s^{s+\frac{1}{n}} H(t,\sigma)d\sigma \cdot x.$$

By (v), for every fixed t the function

$$\sigma \in]a,b[\longmapsto H(t,\sigma)x \in E''_{\sigma(E'',E')}$$

is continuous on the right and hence

$$n\int_s^{s+\frac{1}{n}} H(t,\sigma)d\sigma \cdot x \longrightarrow H(t,s)x$$

when $n \longrightarrow \infty$, which proves that $H(t,s)x = 0$ for every $x \in E$, that is $H(t,s) = 0$.

§ 3 - EXTENSIONS OF THEOREM 2.8

1. Theorem 2.8 may be adapted to the case in which the system

$$L[y] = f, \quad F[y] = 0$$

has one and only one solution for every $f \in C([a,b],E)$. In this case

J_t is a continuous one-to-one linear mapping of E onto

$$E_o = F[C^{(1)}([a,b],E];$$

J_t^{-1} is continuous if and only if E_o is a closed subspace of E. We have $F_t[J(t)^{-1}] = I_{E_o}$ and in the expansion for the Green function G, $\hat{J}(t)^{-1}$ goes from E_o'' onto E''.

2. Instead of a lateral condition $F \in L[C([a,b],E),E]$ it would be more natural to take $\hat{F} \in L[C^{(1)}([a,b],E),E]$. This case can be reduced to the preceding one:

THEOREM 3.1. *If the system* $L[y] = f$, $\hat{F}[y] = \hat{c}$ *has one and only one solution* $y \in C^{(1)}([a,b],E)$ *for every* $f \in C([a,b],E)$ *and* $\hat{c} \in E$, *where* $\hat{F} \in L[C^{(1)}([a,b],E),E]$ *then one can reduce it to a system* $L[y] = f$, $F[y] = c$ *that also has one and only one solution and where*

$$F \in L[C([a,b],E),E].$$

F *and* c *are given by*

$$F[y] = \hat{F}_t\left[y(a) - \int_a^t A(s)y(s)ds\right] \quad \text{and} \quad c = \hat{c} - \hat{F}_t\left[\int_a^t f(s)ds\right].$$

3. Theorem 2.8 may also be extended to systems of the form

$$L[y] \equiv A_o(A_1y)' + By = f, \quad F[y] = c$$

where $A_o, A_1, B \in C([a,b],L(E))$, A_o, A_1 are invertible at every point $t \in [a,b]$. In this case

$$y \in D_L = \{u \in C([a,b],E) \mid A_1u \in C^{(1)}([a,b],E)\};$$

D_L is endowed with the norm

$$\|u\|^{(L)} = \sup[\|u\|, \|(A_1u)'\|]$$

and $F \in L(D_L,E)$.

The systems above can be transformed easily into the ones that appear in Theorem 2.8. In fact, for this purpose we observe that if $L[y] = f$ then

$$y = A_1^{-1}[(A_1y)(a)] - A_1^{-1}I(A_0^{-1}By) + A_1^{-1}I(A_0^{-1}f),$$

where

$$(Ig)(t) = \int_a^t g(s)ds$$

and hence $F[y] = c$ is equivalent to

$$F[A_1^{-1}((A_1y)(a)) - A_1^{-1}I(A_0^{-1}By)] = c - F[A_1^{-1}I(A_0^{-1}f)];$$

if we define

$$F_0[y] = F \circ A_1^{-1} \circ [\delta_{(a)} \circ A_1 - I \circ A_1^{-1} \circ B](y),$$

where $\delta_{(a)}(g) = g(a)$, then the given system is equivalent to the system $L[y] = f$, $F_0[y] = d$ where $d = c - F[A_1^{-1}I(A_0^{-1}f)]$ and now

$$F_0 \in L[C([a,b],E),E].$$

In order to reduce it further to a system of the form

$$L_1[z] = z' + Az = g, \quad F_1[z] = d$$

where $A \in C([a,b],L(E))$, we take $z = A_1y$ and so we obtain the system

$$L_1[z] = z' + Az = g, \quad F_1[z] = d,$$

where $A = A_0^{-1} \circ B \circ A_1^{-1}$, $g = A_0^{-1}f$ and $F_1 = F_0 \circ A_1^{-1}$. It is now easy to verify that if G_1 is the Green function of this last system then the Green function G of the original system $L[y] = f$, $F[y] = c$ is given by $G(t,s) = A_1(t)^{-1} \circ G_1(t,s) \circ A_0(s)^{-1}$.

4. The results of this paper may also be extended to half-open and open intervals, to the case where F takes values in a Banach space different from E, to the case where $A \in L_1^{loc}([a,b],L(E))$ etc. (see [H]).

Note added in proof. The Green function defined in Theorem 2.8, as a function of the first variable satisfies obviously

$$L[y] = 0 \quad \text{for} \quad t \neq s \qquad F[y] = 0.$$

As a function of the second variable it satisfies the non homogeneous integral equation

$$G(t,s) + Y(s-t)R(t,s) - \hat{J}(t)^{-1}\alpha_F(s) = \int_a^s [G(t,\sigma)+Y(\sigma-t)R(t,\sigma)]A(\sigma)d\sigma.$$

Universidade de S.Paulo
Instituto de Matemática e Estatística
Caixa Postal 20570
S.Paulo
BRASIL

REFERENCES

[B] N.BOURBAKI, Fonctions d'une variable réelle, Hermann, Paris, 1951.

[B-K] J.BATT und H.KÖNIG, *Darstellung linearer Transformationen durch Vektorwertige Riemann-Stieltjes-Integrale*, Archiv der Math. X(1959), 273-287.

[C] H.CARTAN, Calcul Différentiel, Hermann, Paris, 1970.

[D] N.DINCULEANU, Vector Measures, Pergamon Press, Oxford, 1967.

[F] J.DIEUDONNÉ, Foundations of Modern Analysis, Academic Press, 1960

[H] C.S.HÖNIG, *The abstract Riemann-Stieltjes integral and its applications to linear differential equations with generalized boundary conditions*, Notas do Instituto de Matemática e Estatística da Universidade de São Paulo, São Paulo, 1973.

[H-P] E.HILLE and R.PHILLIPS, Functional Analysis and Semi-groups, American Mahhematical Society Colloquium Publications, 1957.

SINGULAR PERTURBATIONS OF HYPERBOLIC SYSTEMS

by

Haïm Brezis

The results we discuss here were obtained recently in a joint work with C.Bardos and D.Brezis; complete proofs can be found in [1].

Let $\Omega \subset \mathbb{R}^n$ be a bounded domain with smooth boundary $\partial\Omega$ and outward normal ν. Let $A_i(x)$, $1 \leqslant i \leqslant n$, be mxm symmetric matrices with smooth coefficients; let $k > k_o$ be a constant large enough.

For every $f \in L^2(\Omega)^m$ and for every $\varepsilon > 0$ there exists a unique solution $u_\varepsilon \in H^2(\Omega)^m$ of the equation

(1)
$$-\varepsilon\Delta u_\varepsilon + \sum_{i=1}^{n} A_i \frac{\partial u_\varepsilon}{\partial x_i} + ku_\varepsilon = f \quad \text{on} \quad \Omega$$
$$u_\varepsilon = 0 \quad \text{on} \quad \partial\Omega$$

Indeed, it is sufficient to verify that coerciveness holds,

$$\varepsilon\int_\Omega |\text{grad } u|^2 dx + \int_\Omega \sum_{i=1}^{n} A_i \frac{\partial u}{\partial x_i}\cdot u dx + k\int_\Omega |u|^2 dx \geqslant$$
$$\geqslant \varepsilon\int_\Omega |\text{grad } u|^2 dx + (k-k_o)\int_\Omega |u|^2 dx$$

(note that

$$\int_\Omega A_i \frac{\partial u}{\partial x_i} u dx = -\frac{1}{2}\int_\Omega \frac{\partial A_i}{\partial x_i} u\cdot u dx \ )$$

where k_o depends only on $\sum_{i=1}^{n} \frac{\partial A_i}{\partial x_i}$.

Therefore we get the estimates

(2)
$$\varepsilon\int_\Omega |\text{grad } u_\varepsilon|^2 dx \leqslant C$$

(3) $$\int_\Omega |u_\varepsilon|^2 dx \leqslant C$$

where C is independent of ε. Consequently there is a sequence $\varepsilon_n \longrightarrow 0$ such that u_{ε_n} converges weakly in $L^2(\Omega)^m$ to u; clearly u is a solution in the sense of distributions of the equation

(4) $$\sum_{i=1}^{n} A_i \frac{\partial u}{\partial x_i} + ku = f.$$

Our purpose is to answer the following question raised by J.L.Lions.

PROBLEM: Identify the boundary conditions that u satisfies on $\partial\Omega$.

In the case where $m = 1$ (i.e. one single equation) the answer has been known for a long time (see e.g. [1]). We recall it briefly here: let $\nu A(x)$ be the function defined on $\partial\Omega$ by

$$\nu A(x) = \sum_{i=1}^{n} \cos(\nu, x_i) A_i(x)$$

and let

$$\Sigma_+ \text{ (resp. } \Sigma_-) = \{x \in \partial\Omega \ ; \ \nu A(x) > 0 \ \text{ (resp. } \nu A(x) < 0)\}.$$

Then, under same assumptions (the boundaries of Σ_+ and Σ_- in $\partial\Omega$ should be smooth n-2 dimensional manifolds) one has the following

THEOREM 0. *Assume* $m = 1$; *as* $\varepsilon \longrightarrow 0$ *the solution* u_ε *of* (1) *converges in* $L^2(\Omega)$ *to* u *which satisfies* (4) *and*

(5) $$u = 0 \quad on \quad \Sigma_- .$$

In addition, the problem (4)-(5) is well posed (in an appropriate weak sense).

The proof of Theorem 0 which is given in [1] relies heavily on the maximum principle and cannot be used when Δ is replaced by a higher order elliptic operator; also it does not extend to systems $(m > 1)$.

The following results for systems are proved by completely different techniques (energy methods). Let us now consider the *matrix* $\nu A(x)$ defined for $x \in \partial\Omega$ by

$$\nu A(x) = \sum_{i=2}^{n} \cos(\nu, x_i) A_i(x).$$

Recall the following definition: given an $m \times m$ matrix B and a subspace $N \subset \mathbb{R}^m$ we say that N is *maximal positive* if $Bz \cdot z \geqslant 0$ for every $z \in N$ and for each subspace $N' \supset N$, $N' \neq N$, there is some $z' \in N'$ such that $Bz' \cdot z' < 0$. (Note that, in general, there are many maximal positive spaces for a given B).

The results of Friedrichs, Lax, Phillips assert that if $N(x)$ is maximal positive for $\nu A(x)$ $(x \in \partial\Omega)$ and if $N(x)$ depends smoothly on $x \in \partial\Omega$, then for every $f \in L^2(\Omega)^m$ the problem

$$(6) \qquad \begin{cases} \displaystyle\sum_{i=1}^{n} A_i \frac{\partial u}{\partial x_i} + ku = f & \text{on} \quad \Omega \\ u(x) \in N(x) & \text{on} \quad \partial\Omega \end{cases}$$

is well posed (in an appropriate weak sense).

NOTATION. Given an $m\times m$ symmetric matrix B we denote by B^+ the space generated by all eigenvectors of B corresponding to nonnegative eigenvalues of B. Clearly B^+ is maximal positive for B.

THEOREM 1. *Assume* $(\nu A(x))^+$ *depends smoothly on* $x \in \partial\Omega$. *Then, as* $\varepsilon \longrightarrow 0$, *the solution* u_ε *of (1) converges weakly in* $L^2(\Omega)^m$ *to* u, *which satisfies (4) and the boundary condition*

$$u(x) \in (\nu A(x))^+ \quad \text{on} \quad \partial\Omega. \tag{7}$$

REMARKS. 1) It is not known whether u_ε converges strongly in $L^2(\Omega)^m$.

2) The conclusion of Theorem 1 can actually be "localized" on the parts of $\partial\Omega$ where $(\nu A(x))^+$ is smooth.

Sketch of the proof of Theorem 1

We stretch $\partial\Omega$ after localization; so that (1) becomes

$$-\varepsilon L u_\varepsilon + \sum_{i=1}^{n} A_i \frac{\partial u_\varepsilon}{\partial x_i} + k u_\varepsilon = f \quad \text{on} \quad \mathbb{R}^n_+ \tag{8}$$

where

$$L = \sum_{i,j=1}^{n} a_{ij} \frac{\partial^2}{\partial x_i \partial x_j} + \sum_{i=1}^{n} a_i \frac{\partial}{\partial x_i},$$

and u_ε has compact support.

In the new coordinates $\nu A = -A_n$.

Equation (8) can be written as

$$-\varepsilon a_{nn} \frac{\partial^2 u_\varepsilon}{\partial x_n^2} + A_n \frac{\partial u_\varepsilon}{\partial x_n} = g_\varepsilon \quad \text{on} \quad \mathbb{R}^n_+ \tag{9}$$

where

$$g_\varepsilon = f - ku_\varepsilon - \sum_{i=1}^{n-1} A_i \frac{\partial u_\varepsilon}{\partial x_i} +$$

$$+ \varepsilon \sum_{(i,j)\neq(n,n)} a_{ij} \frac{\partial^2 u_\varepsilon}{\partial x_i \partial x_j} + \varepsilon \sum_{i=1}^{n} a_i \frac{\partial u_\varepsilon}{\partial x_i}$$

and by (2)-(3) g_ε remains bounded in the space $L^2(\mathbb{R}_+ ; H^{-1}(\mathbb{R}^{n-1}))^m$.

Changing the coordinates (in a smooth way) we can assume that A_n takes the form

$$A_n = \begin{pmatrix} A_n^1 & 0 \\ 0 & A_n^2 \end{pmatrix}$$

where A_n^1 is an $r\times r$ positive definite matrix and $-A_n^2$ is nonnegative. Let us denote by $\bar{u}_\varepsilon$ the first r components of u_ε. So that $\bar{u}_\varepsilon$ satisfies

$$-\varepsilon \frac{\partial^2 \bar{u}_\varepsilon}{\partial x_n^2} + P \frac{\partial \bar{u}_\varepsilon}{\partial x_n} = \bar{g}_\varepsilon \tag{10}$$

where

$$P(x)z\cdot z \geqslant \delta|z|^2 \qquad (\delta > 0)$$

and $\bar{g}_\varepsilon$ is bounded in $L^2(\mathbb{R}_+ ; H^{-1}(\mathbb{R}^{n-1}))^r$. Let

$$\Delta' = \sum_{i=1}^{n-1} \frac{\partial^2}{\partial x_i^2}$$

and

$$v_\varepsilon = (\mu - \Delta')^{-1} \frac{\partial \bar{u}_\varepsilon}{\partial x_n}, \qquad h_\varepsilon = (\mu - \Delta')^{-1} \bar{g}_\varepsilon$$

where $\mu > 0$.

Hence (10) becomes

$$(11) \qquad -\varepsilon(\mu - \Delta')\frac{\partial v_\varepsilon}{\partial x_n} + P(\mu - \Delta')v_\varepsilon = (\mu - \Delta')h_\varepsilon$$

and h_ε is bounded in $L^2(\mathbb{R}_+ ; H^1(\mathbb{R}^{n-1}))^r$. Multiplying (11) through by v_ε and integrating on $\mathbb{R}^n_+$ we get

$$\int_{\mathbb{R}^{n-1}} \frac{\varepsilon\mu}{2}|v_\varepsilon(x',0)|^2 dx' + \frac{\varepsilon}{2}\int_{\mathbb{R}^{n-1}} \sum_{i=1}^{n-1}\left|\frac{\partial v_\varepsilon}{\partial x_i}(x',0)\right|^2 dx' +$$

$$+ \mu\delta\int_{\mathbb{R}^n_+}|v_\varepsilon|^2 dx + \delta\int_{\mathbb{R}^n_+}\sum_{i=1}^{n-1}\left|\frac{\partial v_\varepsilon}{\partial x_i}\right|^2 dx +$$

$$+ \int_{\mathbb{R}^n_+}\sum_{i=1}^{n-1}\frac{\partial P}{\partial x_i}\frac{\partial v_\varepsilon}{\partial x_i} v_\varepsilon dx \leqslant \mu\int_{\mathbb{R}^n_+} h_\varepsilon v_\varepsilon dx + \int_{\mathbb{R}^n_+}\sum_{i=1}^{n-1}\frac{\partial h_\varepsilon}{\partial x_i}\frac{\partial v_\varepsilon}{\partial x_i} dx.$$

By choosing μ large enough we can "absorb"

$$\frac{\partial P}{\partial x_i}\frac{\partial v_\varepsilon}{\partial x_i} v_\varepsilon \qquad \text{into} \qquad \delta\left|\frac{\partial v_\varepsilon}{\partial x_i}\right|^2 + \mu\delta|v_\varepsilon|^2 .$$

Therefore v_ε remains bounded in $L^2(\mathbb{R}_+ ; H^1(\mathbb{R}^{n-1}))^r$ and $\frac{\partial \bar{u}_\varepsilon}{\partial x_n}$ remains bounded in $L^2(\mathbb{R}_+ ; H^{-1}(\mathbb{R}^{n-1}))^r$. This is sufficient to conclude that, at limit, $\bar{u} = 0$ on $\partial\Omega$, which is equivalent to the fact that $u(x) \in (\nu A(x))^+$ on $\partial\Omega$.

Theorem 1 still holds true when Δ is replaced by any elliptic operator

$$L = \sum_{|\alpha| \leqslant 2s} a_\alpha D^\alpha$$

of order 2s (note that L is *not an elliptic system* but one *single* elliptic operator acting similarly on each component of u).

The previous proof can be modified as follows. First of all, (3) is still valid, but (2) is replaced by (Gårding's inequality)

(12) $$\varepsilon \|u_\varepsilon\|^2_{H^s} \leqslant C.$$

Using interpolation, (3) and (12) we get

(13) $$\|u_\varepsilon\|^2_{H^1} \leqslant C\ \varepsilon^{-\frac{1}{2s}}.$$

Applying a standard regularity theorem to the equation, $u_\varepsilon \in H^s_o(\Omega)^m$,

$$\varepsilon L u_\varepsilon + \sum_{i=1}^{n} A_i \frac{\partial u_\varepsilon}{\partial x_i} + k u_\varepsilon = f$$

we get

(14) $$\varepsilon \|u_\varepsilon\|_{H^{2s}} \leqslant C_1 + C_2 \|u_\varepsilon\|_{H^1} \leqslant C \varepsilon^{-\frac{1}{2s}} \qquad (\varepsilon < 1).$$

Interpolating again between (12) and (14) we obtain

(15) $$\|u_\varepsilon\|_{H^{2s-1}} \leqslant C\varepsilon^{-1+\frac{1}{2s^2}}$$

Next (10) is replaced

$$\varepsilon(-1)^s \frac{\partial^{2s}\bar{u}_\varepsilon}{\partial x_n^{2s}} + P\frac{\partial \bar{u}_\varepsilon}{\partial x_n} = \bar{g}_\varepsilon$$

where $\bar{g}_\varepsilon$ is bounded in $L^2(\mathbb{R}_+ ; H^{-1}(\mathbb{R}^{n-1}))^r$ (use (15)). We conclude as in the previous proof that $\frac{\partial \bar{u}_\varepsilon}{\partial x_n}$ is bounded in $L^2(\mathbb{R}_+ ; H^{-1}(\mathbb{R}^{n-1}))^r$.

In case Δ is replaced by an elliptic *system* E of order 2s the situation is more complicated. Let

$$E = \sum_{|\alpha|\leq 2s} E_\alpha D^\alpha$$

be an elliptic system and suppose that the E_α's are symmetric matrices for $|\alpha| = 2s$. Let $\nu E(x)$ be the matrix defined for $x \in \partial\Omega$ by

$$\nu E(x) = (-1)^s \sum_{|\alpha|=2s} \nu^\alpha E_\alpha(x)$$

where $\nu^\alpha = \cos(\nu,x_1)^{\alpha_1}\cos(\nu,x_2)^{\alpha_2}\ldots\cos(\nu,x_n)^{\alpha_n}$. νE is symmetric and positive definite (since E is elliptic).

Let us denote by $N(x)$ the space generated by all eigenvectors of $\nu A(x)$ relative to $\nu E(x)$, which correspond to nonnegative eigenvalues (i.e. we consider the equations $(\nu A)z = \lambda(\nu E)z$, $\lambda \geq 0$). In other words

$$N = (\nu E)^{\frac{1}{2}} \left[(\nu E)^{-\frac{1}{2}}(\nu A)(\nu E)^{-\frac{1}{2}}\right]^+ .$$

Clearly $N(x)$ is maximal positive for $\nu A(x)$.

THEOREM 2. *Assume* $N(x)$ *depends smoothly on* $x \in \partial\Omega$. *Then as* $\varepsilon \longrightarrow 0$, *the solution* $u_\varepsilon \in H_o^s(\Omega)^m$ *of*

$$\varepsilon E u_\varepsilon + \sum_{i=1}^n A_i \frac{\partial u_\varepsilon}{\partial x_i} + k u_\varepsilon = f \quad \text{on} \quad \Omega$$

converges weakly in $L^2(\Omega)^m$ *to* u *which satisfies* (4) *and the boundary condition*

$$u(x) \in N(x) \quad on \quad \partial\Omega.$$

Sketch of the proof. Instead of (9) we have now

$$\text{(16)} \qquad \varepsilon(-1)^s (\nu E) \frac{\partial^{2s} u_\varepsilon}{\partial x_n^{2s}} + A_n \frac{\partial u_\varepsilon}{\partial x_n} = g_\varepsilon \quad \text{on} \quad \mathbb{R}^n_+$$

where g_ε is bounded in $L^2(\mathbb{R}_+ ; H^{-1}(\mathbb{R}^{n-1}))^n$. Let $v_\varepsilon = (\nu E)^{\frac{1}{2}} u_\varepsilon$, so that, after throwing away a number of commutators, (16) becomes

$$\text{(17)} \qquad (-1)^s \frac{\partial^{2s} v_\varepsilon}{\partial x_n^{2s}} + (\nu E)^{-\frac{1}{2}} A_n (\nu E)^{-\frac{1}{2}} \frac{\partial v_\varepsilon}{\partial x_n} = \tilde{g}_\varepsilon$$

where $\tilde{g}_\varepsilon$ is bounded in $L^2(\mathbb{R}_+ ; H^{-1}(\mathbb{R}^{n-1}))^m$. We are now reduced to the previous situation where A_n is replaced by

$$(\nu E)^{-\frac{1}{2}} A_n (\nu E)^{-\frac{1}{2}}.$$

Hence we know that, at the limit, $v(x) \in \left[(\nu E)^{-\frac{1}{2}} (\nu A)(\nu E)^{-\frac{1}{2}}\right]^+$ on $\partial\Omega$ and therefore $u(x) \in N(x)$ on $\partial\Omega$.

Université de Paris VI
Département de Mathématiques
9 Quai St. Bernard
Paris V^e
FRANCE

REFERENCES

[1] C.BARDOS, D.BREZIS, H.BREZIS, *Perturbations singulières et prolongements maximaux d'opérateurs positifs* (to appear, Archive Rat. Mech. Anal.).

[2] C.BARDOS, *Problèmes aux limites pour les équations du premier ordre*, Ann. Sc. E.N.S. 4eme Série, t.3(1970), pp. 185-233.

THE BOUNDED CASE OF THE WEIGHTED APPROXIMATION PROBLEM

by

W.H. Summers [1]

Since recent developments make it possible to give an essentially complete treatment of that aspect of the weighted approximation problem known as the bounded case, we take this opportunity to present a brief exposition on this topic. We will then mention a related, but still open, question for which the situation in the bounded case may well provide the necessary key.

1. THE WEIGHTED APPROXIMATION PROBLEM. We introduce a set V of nonnegative upper semicontinuous (*u.s.c.*) functions defined on a completely regular Hausdorff space X; the elements of V being referred as *weights*. The corresponding *weighted space* $CV_o(X)$ is the locally convex topological vector space obtained by equipping the vector space consisting of those $f \in C(X)$, the complex valued continuous functions on X, such that fv vanishes at infinity for every $v \in V$ with the *weighted topology* ω_V generated by the seminorms $f \longmapsto \|fv\|$, one for each $v \in V$, where $\|\cdot\|$ denotes the usual supremum norm defined on the bounded functions on X. Since there is no loss in generality, we will assume that if $u, v \in V$ and $\lambda \geq 0$, then there is a $w \in V$ for which $\lambda u, \lambda v \leq w$ (pointwise); i.e., V is a Nachbin family on X. (Several examples together with a discussion of weighted spaces as a class of locally convex spaces are given in our recent expository article [9].) Furthermore, we hereafter assume that a subalgebra A of $C(X)$ and a linear subspace W of $CV_o(X($, where W is an A-module with respect to pointwise multiplication, have been specified.

Formulated and studied by Nachbin [3], [4], [5], the *weighted approximation problem*, a version of the classical Bernstein approximation problem, asks for a description of the closure of W in $CV_o(X)$. In particular, motivated by Stone's treatment of the Weierstrass approxi-

1 Supported in part by National Science Foundation Grant GP-34370

mation theorem, Nachbin was concerned with when $c\ell(W)$ could be described in terms of a localization (to be discussed in the following section) of the weighted approximation problem, and he established, assuming A to be selfadjoint, that this description is valid in certain cases. One of these, the one in which every $a \in A$ is bounded on the support of each $v \in V$, is termed the *bounded case* of the weighted approximation problem. This case was introduced by Nachbin [5, p.294] in 1965 as a bridge to more general results, but more about this later. We note in passing that instances of the bounded case occur whenever A consists only of bounded functions or each $v \in V$ has compact support.

We should mention, before continuing, that the weighted approximation problem has also been considered in other settings; for example, see the articles by Prolla [8] and Nachbin, Machado, and Prolla [7].

2. LOCALIZATION OF THE WEIGHTED APPROXIMATION PROBLEM. If K is a closed subset of X, then the set $V|K$ of restrictions to K of all functions in V is a Nachbin family on K, the set $W|K$ of restrictions to K of all elements in W is a linear subspace of the weighted space $C(V|K)_o(K)$, and $W|K$ is an $A|K$-module, where $A|K$ denotes the algebra of restrictions to K of all members of A. Consequently, if $\mathcal{K}$ is a pairwise disjoint closed covering of X, then we have a *localization* of the weighted approximation problem (relative to $\mathcal{K}$) provided $c\ell(W)$ consists precisely of those $f \in CV_o(X)$ such that the restriction of f to K belongs to the closure of $W|K$ in $C(V|K)_o(K)$, for every $K \in \mathcal{K}$.

In his approach to the weighted approximation problem, Nachbin was interested in localization relative to the collection of equivalence classes corresponding to the equivalence relation defined on X by considering $x, y \in X$ to be equivalent whenever $a(x) = a(y)$ for every $a \in A$, and he refers to W as being *localizable under* A [4, p.126] in those instances where it is possible to achieve this particular localization. In this connection, Nachbin established the following result relating to the bounded case.

2.1 THEOREM (Nachbin [5, p.295]). *If* A *is selfadjoint, then* W *is localizable under* A *in the bounded case of the weighted approximation problem.*

3. THE GENERAL COMPLEX CASE. What can be said when A is not assumed to be selfadjoint? Interest in this question dates from the formative stages of the weighted approximation problem (cf. [3], [4]), but not until recently has much progress been realized. We now proceed to discuss these recent developments.

It is easy to see that the bounded case alone is not sufficient for W to be localizable under A. Indeed, if we take X to be the complex plane, V to be the set of all nonnegative continuous functions on X which have compact support, A to be the algebra of entire functions, and W = A, then we have an instance of the bounded case where W is clearly not localizable under A (otherwise, W would be compact-open dense in C(X)). Consequently, unless restrictions are placed on A, the only hope for extending Nachbin's theory for the selfadjoint bounded case lies in considering a localization of the weighted approximation problem relative to a pairwise disjoint closed covering of X which will agree with the one studied by Nachbin whenever A is selfadjoint. A possible candidate is suggested by Bishop's treatment [1] of the Stone-Weierstrass theorem.

A set $K \subseteq X$ is called an *antisymmetric set* for A if each $a \in A$ which is real valued on K is necessarily constant on K. It readily follows that every antisymmetric set for A is contained in a maximal antisymmetric set for A, and that the collection $\mathcal{K}_A$ of maximal antisymmetric sets for A is a pairwise disjoint closed covering of X which coincides with the one defined in Section 2 whenever A is selfadjoint. We will say that W is A-*localizable* if localization occurs relative to $\mathcal{K}_A$.

One step in the development of the bounded case was provided by J.B.Prolla. We write $V \leqslant C^+(X)$ if, for each $v \in V$, there is a corresponding $f \in C(X)$ such that $v \leqslant f$.

3.1. THEOREM (Prolla [8, p.284]). *Assume* X *is locally compact. If* $V \leqslant C^+(X)$, *then* W *is* A-*localizable in the bounded case of the weighted approximation problem.*

At about the same time, we obtained a related result concerning $(C_b(X), \beta_o)$, the space of bounded continuous functions on X equipped with the substrict topology.

3.2. THEOREM ([10, p.92]). *If* V *is the Nachbin family of all nonnegative u.s.c. functions on* X *which vanish at infinity, then* W *is* A-*localizable in the bounded case of the weighted approximation problem.*

Neither of the two theorems above contain Nachbin's theorem for the selfadjoint case. However, it is possible to apply Theorem 3.2 in order to obtain a generalization of Theorem 2.1. We will let S(A) denote the subalgebra of all $f \in C(X)$ which are constant on each $K \in \mathcal{K}_A$.

3.3. THEOREM ([10, p.96]). *If* $\mathcal{K}_{S(A)} = \mathcal{K}_A$ *, then* W *is* A-*localizable in the bounded case of the weighted approximation problem.*

Although replacing the selfadjointness hypothesis by the condition $\mathcal{K}_{S(A)} = \mathcal{K}_A$ was only a small step, it led to the conjecture that all restrictions on A could be removed [10, p.96], and this has just recently been confirmed.

3.4. THEOREM ([11]). *In the bounded case of the weighted approximation problem,* W *is always* A-*localizable.*

4. SOME CONSEQUENCES. Theorem 3.4 can be regarded as an extension of Bishop's generalized Stone-Weierstrass theorem [1] to the class of weighted spaces under consideration. Indeed, appropriate choices of X and V yield the setting for both Bishop's theorem and Glicksberg's analogous result [2] for the strict topology. At the same time, of course, many noteworthy versions of the classical Stone-Weierstrass theorem are subsumed (see [10]). In fact, an exceedingly general result of Stone-Weierstrass type can be deduced from Theorem 3.4 (or even Theorem 3.3). This occurs when each $K \in \mathcal{K}_A$ consists of just a single point; i.e., in the *separating case* of the weighted approximation problem.

4.1. THEOREM ([10, p.97]). *In the separating and bounded case of the weighted approximation problem,* W *is dense in* $CV_o(X)$ *if and only if, corresponding to each* $x \in X$ *for which there exist* $v \in V$ *and* $f \in CV_o(X)$ *such that both* $v(x) \neq 0$ *and* $f(x) \neq 0$, *there is a* $w \in W$ *with* $w(x) \neq 0$.

5. AN OPEN QUESTION. Assuming A to be selfadjoint, Nachbin [5] was able to use Theorem 2.1 in order to establish a more general criterion for localizability under A. This case, termed the analytic criterion [6, p.91], is the one in which there exist subsets $G(A)$ of A and $G(W)$ of W such that the following conditions are satisfied:

(1) the subalgebra of A generated by $G(A)$ is compact-open dense in A;

(2) the A-submodule of W generated by $G(W)$ is ω_V-dense in S;

and

(3) if $v \in V$, $a \in G(A)$, and $w \in G(W)$, the positive numbers α and β exist so that

$$|w(x)|v(x) \leqslant \alpha e^{-\beta |a(X)|}$$

holds for all $x \in X$.

The three conditions listed above are clearly satisfied by A and W in the bounded case of the weighted approximation problem. Therefore, this situation, which we will call the *analytic case* of the weighted approximation problem, may possibly provide a more general criterion for localizability even when A is not assumed to be selfadjoint. In particular, is W necessarily A-localizable in the analytic case of the weighted approximation problem?

University of Arkansas
Department of Mathematics
Fayetteville, Ark. 72701
U.S.A.

REFERENCES

[1] E.BISHOP, *A generalization of the Stone-Weierstrass theorem*, Pacific J.Math. 11(1961), 777-783.

[2] I.GLICKSBERG, *Bishop's generalized Stone-Weierstrass theorem for the strict topology*, Proc. Amer. Math. Soc. **14**(1963), 329-333.

[3] L.NACHBIN, *On the weighted polynomial approximation in a locally compact space*, Proc. Nat. Acad. Sci. **47**(1961),1055-1057.

[4] L.NACHBIN, *Weighted approximation over topological spaces and the Bernstein problem over finite dimensional vector spaces*, Topology 3, Suppl. 1(1964), 125-130.

[5] L.NACHBIN, *Weighted approximation for algebras and modules of continuous functions: real and self-adjoint complex cases*, Ann. of Math. **81**(1965), 289-302.

[6] L.NACHBIN, *Elements of approximation theory*, Van Nostrand, Princeton, NJ, 1967.

[7] L.NACHBIN, S.MACHADO, and J.B.PROLLA, *Weighted approximation, vector fibrations, and algebras of operators*, J.Math. Pures et Appl. **50**(1971), 299-323.

[8] J.B.PROLLA, *Bishop's generalized Stone-Weierstrass theorem for weighted spaces*, Math. Ann. **191**(1971), 283-289.

[9] W.H.SUMMERS, *Weighted spaces and weighted approximation*, Séminaire d'Analyse Moderne, Université de Sherbrooke, Sherbrooke, PQ, Nº 3 (1970).

[10] W.H.SUMMERS, *The general complex bounded case of the strict weighted approximation problem*, Math. Ann. **192**(1971), 90-98.

[11] W.SUMMERS, *Weighted approximation for modules of continuous functions*, Bull. Amer. Math. Soc. **79** (1973), 386-388.

ON NONLINEAR INTEGRAL EQUATIONS OF HAMMERSTEIN TYPE WITH UNBOUNDED LINEAR MAPPING *

by

Chaitan P. Gupta

INTRODUCTION

Let Ω be a measurable subset of the n-dimensional Euclidean space $\mathbb{R}^n$ and let X and Y denote Banach spaces of real-valued measurable functions on Ω. A real-valued function $f(x,t)$ defined on the cartesian product $\Omega \times \mathbb{R}$ is said to satisfy Caratheodory's conditions if (i) $f(x,t)$ is a measurable function of $x \in \Omega$ for each $t \in \mathbb{R}$ and (ii) for almost all $x \in \Omega$ the function $f(x,t)$ is continuous in t. It is easy to show using Lusin's theorem that $f(x,u(x))$ is a measurable function of $x \in \Omega$ whenever $f(x,t)$ satisfies Caratheodory's conditions and $u(x)$ is a measurable function of x in Ω. We denote by N the mapping defined by $Nu(x) = f(x,u(x))$ from the Banach space X into the Banach space Y. The mapping N is called a Nemytskii mapping. It is well-known that in case $X = L_p(\Omega)$, $Y = L_q(\Omega)$, $1 \leqslant p \leqslant \infty$, $1 \leqslant q \leqslant \infty$ and if a Nemytskii mapping N is well-defined from $L_p(\Omega)$ into $L_q(\Omega)$ then N is a bounded continuous mapping. Further N is well-defined from $L_p(\Omega)$ into $L_q(\Omega)$ if and only if there exists a function $a(x) \in L_q(\Omega)$ and a constant $b > 0$ such that

$$|f(x,t)| \leqslant a(x) + b|t|^{p-1}$$

for $x \in \Omega$ and $t \in \mathbb{R}$. Moreover, a Nemytskii mapping N is compact if and only if N is a constant mapping. (See Krasnoselskii [20] and Vainberg [27] for details.)

A nonlinear integral equation of Hammerstein type in the Banach space X is an equation of the form

---

* This is the text of an hour address given by the author at the Symposium of Analysis held at Recife, Brazil from July 9-29, 1972

(1) $$u(x) + \int_{\Omega} K(x,y) f(y,u(y))\,dy = v(x)$$

for a given funciion $v(x)$ in X and unknown function $u(x)$ in X. Here $K(x,y)$ is a real-valued measurable function on $\Omega \times \Omega$ such that the linear mapping A defined by

$$Aw(x) = \int_{\Omega} K(x,y) w(y)\,dy$$

is well-defined from the Banach space Y into the Banach space X. In abstract form equation (1) can be written, for a given v in X, as

(2) $$u + ANu = v$$

where A is a linear mapping from Y into X and N is a (nonlinear) Nemytskii mapping from X into Y.

Nonlinear integral equations of Hammerstein type have been an object of intense study since the appearance of the celebrated paper of A.Hammerstein ([14]) in Acta Mathematicain 1930. Classical results on nonlinear equations of Harmmerstein type center around the assumptions that A is a bounded linear mapping and N a bounded continuous mapping which is noncompact. The case of compact N being uninteresting for, as observed above, in the case of L_p-spaces N is compact if and only if it is a constant mapping. Further it was always additionally assumed that either the linear mapping A is compact or that the Nemytskii mapping N satisfied a Lipschitz condition which condition could easily be verified if the function $f: \Omega \times \mathbb{R} \longrightarrow \mathbb{R}$ satisfied a condition of the form

$$|f(x,t_1) - f(x,t_2)| \leqslant a(x)|t_1 - t_2|$$

for some $a(x) \in L_q(\Omega)$ and $t_1, t_2 \in \mathbb{R}$. These assumptions lead one to use the method of Leray-Schauder degree theory for mappings of the form $I + C$ (where C is a compact mapping in a Banach space X and I denotes the identity mapping in X) when A is compact and to use Picard's method of iteration for contraction mappings when N satisfies a Lipschitz condition. With the development of the theory of monotone operators from a Banach space X to its dual Banach space X^* by Browder and Minty in the early 1960's a new method to attack the problem of existence of solutions of nonlinear integral equations of Hammerstein type became available. This allowed one to consider the case when neither the linear mapping A was compact nor the nonlinear mapping N satisfied a Lipschitz condition under some monotonicity hypothesis on both A and N. For results on nonlinear equations of Hammerstein type when A is a bounded linear mapping using monotonicity methods we refer the reader to a recent excellent survey paper of Browder [6] and the bibliography there. The case of nonlinear equations of Hammerstein type in a Hilbert space when A is an unbounded linear mapping was first studied by Lavrentiev in a Doklady note [21] in 1966 and has since been further studied by Lavrentiev [22], Vainberg-Lavrentiev [29], Browder-de Figueiredo-Gupta [8], Koscikii [19] and de Figueiredo-Gupta [9], [10], [11]. The purpose of this paper is to present a survey of recent results on nonlinear integral equations of Hammerstein type both in Hilbert and Banach spaces when A is an unbounded linear mapping.

In section 1 we present notations, definitions and some of the preliminary results that we need. In section 2 we study equation (2) in a Banach space and in section 3 we study equation (2) in a Hilbert space.

§1. NOTATIONS, DEFINITIONS AND SOME PRELIMINARY RESULTS

Let X be a real Banach space and X^* denote the dual Banach space of X. We denote by (w,x) the duality pairing between the elements w in X^* and x in X. A subset G of the cartesian product $X \times X^*$ is said to be *monotone* if $(w_1-w_2, x_1-x_2) \geqslant 0$ for $[x_1,w_1] \in G$ and $[x_2,w_2] \in G$ and G is said to be *maximal monotone* if it is not a proper subset of another monotone subset of $X \times X^*$. Let T be a mapping from X into 2^{X^*}, the set of subsets of X^*. The *effective domain* $D(T)$ of T is defined to be the subset of X given by

$$D(T) = \{x \in X \mid Tx \neq \emptyset\}$$

and the *graph* $G(T)$ of T is defined to be the subset of $X \times X^*$ given by $G(T) = \{[x,w] \mid w \in Tx\}$. T is said to be *monotone* if $G(T)$ is a monotone subset of $X \times X^*$ and is said to be *maximal monotone* if $G(T)$ is a maximal monotone subset of $X \times X^*$. A single-valued mapping from X into X^* is said to be *hemi-continuous* if T is continuous from line segments in X to X^* endowed with weak topology. T is said to be *demi-continuous* if it is continuous from the norm topology of X to X^* endowed with weak topology. The following proposition due to Kato [15] shows that hemi-continuous monotone mappings from X into X^* are demi-continuous.

PROPOSITION 1. *Let* T *be a single-valued monotone mapping from a real Banach space* X *into* X^*. *Then* T *is hemi-continuous if and only if it is demi-continuous.*

PROOF. Clearly if T is demi-continuous it is hemi-continuous. Suppose, now, that T is hemi-continuous and we shall show that T is demi-continuous. We first assert that T is locally bounded. Indeed, suppose that T is not locally bounded at a point $u \in X$ so that there exists a sequence $\{u_n\}$ in X with $u_n \longrightarrow u$ and $r_n = \|Tu_n\| \longrightarrow \infty$ as $n \longrightarrow \infty$. Let

$$t_n = \max\{\frac{1}{r_n} , \|u_n-u\|^{\frac{1}{2}}\}$$

so that $r_n t_n \geqslant 1$, $\|u_n-u\| \leqslant t_n^2$ for all n and $t_n \longrightarrow 0$ as $n \longrightarrow \infty$. Now, for $v \in X$ let $w_n = u + t_n v$ so that $w_n - u_n = u - u_n + t_n v$ and

$$\|\frac{1}{t_n}(w_n-u_n)-v\| = \frac{1}{t_n}\|u-u_n\| \leqslant t_n \longrightarrow 0$$

as $n \longrightarrow \infty$. Thus $\frac{1}{t_n}(w_n-u_n) \longrightarrow v$ and also since T is hemi-continuous we have $Tw_n \longrightarrow Tu$ (weakly). Using, now, the monotonicity of T we obtain

$$(Tu_n , v) \leqslant \frac{1}{t_n}(Tu_n , u_n-u) - (Tw_n , \frac{1}{t_n}(u_n-w_n))$$

for all n. It then follows that there exists a constant M_v (depending on v) such that

$$\left|(\frac{1}{r_n t_n} Tu_n , v)\right| \leqslant M_v$$

for all n. But this contradicts the principle of uniform boundedness since

$$\frac{1}{r_n t_n}\|Tu_n\| = \frac{1}{t_n} \longrightarrow +\infty$$

as $n \longrightarrow \infty$. Thus T is locally bounded on X. Now to prove that T is demi-continuous let $u_n \longrightarrow u$ in X so that $\{\|Tu_n\|\}$ is bounded. Letting $t_n = \|u_n - u\|^{1/2}$ and for $v \in X$, $w_n = u + t_n v$, we obtain as above that $\frac{1}{t_n}(w_n - u_n) \longrightarrow v$, $Tw_n \rightharpoonup Tu$ (weakly) and

$$(Tu_n , v) \leqslant \frac{1}{t_n}(Tu_n , u_n - u) - (Tw_n , \frac{1}{t_n}(u_n - w_n))$$

for all n. It then, follows from the boundedness of $\{\|Tu_n\|\}$ and $(Tw_n , \frac{1}{t_n}(w_n - u_n)) \longrightarrow (Tu,v)$ that

$$\begin{aligned} |(Tu_n ,v) - (Tu,v)| &\leqslant \frac{1}{t_n}\|Tu_n\| \, \|u_n - u\| + |(Tw_n , \frac{1}{t_n}(w_n - u_n)) - (Tu,v)| \\ &= \|Tu_n\| \, \|u_n - u\|^{\frac{1}{2}} + |(Tw_n , \frac{1}{t_n}(w_n - u_n)) - (Tu,v)| \\ &\longrightarrow 0 \quad \text{as} \quad n \longrightarrow \infty. \end{aligned}$$

Hence $Tu_n \rightharpoonup Tu$ (weakly) and so T is demi-continuous. This completes the proof of the Proposition. Q.E.D.

DEFINITION 1. A single-valued mapping T from X into X^* is said to be of type (M) if the following two conditions hold:

(M_1) For any sequence $\{u_n\}$ in X which converges weakly to an element u_o in X, the sequence $\{Tu_n\}$ converges weakly to an element w_o in X^* and $\lim\sup(Tu_n , u_n) \leqslant (w_o, x_o)$ then $Tu_o = w_o$.

(M_2) T is continuous from finite dimensional subspaces of X to X^* endowed with weak topology.

The concept of mappings of type (M) was first introduced by Brézis in [1] using filters instead of sequences as we do above. We may note that our definition of a mapping of type (M) is equivalent to that introduced by Brézis if the Banach space X is separable and the mapping T in question is bounded. We need the following results from [1], [9], [23] on mappings of type (M) and remark that a weakly sequentially continuous mapping from X into X^* is of type (M).

PROPOSITION 2. *Let* X *be a real Banach space and* $T: X \longrightarrow X^*$ *a monotone hemi-continuous mapping. Then* T *is a mapping of type* (M).

PROOF. Since it follows from Proposition 1 that T is a demi-continuous mapping, T satisfies condition (M_2) in Definition 1. Now to verify that T satisfies (M_1) let $\{u_n\}$ be a sequence in X such that $u_n \longrightarrow u_0 \in X$ (weakly), $Tu_n \longrightarrow w_0 \in X^*$ (weakly) and

$$\limsup (Tu_n , u_n) \leqslant (w_0 , u_0).$$

Now for $u \in X$ we have using the monotonicity of T that

$$0 \leqslant (Tu-Tu_n , u-u_n) = (Tu , u-u_n) - (Tu_n , u) + (Tu_n , u_n).$$

This gives on taking limits as $n \to \infty$ that

$$\begin{aligned} 0 &\leqslant (Tu , u-u_0) - (w_0 , u) + \liminf (Tu_n , u_n). \\ &\leqslant (Tu , u-u_0) - (w_0 , u) + (w_0 , u_0) \\ &= (Tu-w_0 , u-u_o) \end{aligned}$$

for $u \in X$. It is then easy to see using the hemi-continuity of T that

$Tu_o = v_o$. Thus T satisfies (M_1) and the proof of the proposition is complete. Q.E.D.

It is easy to see that the sum of two mappings of type (M) may not be a mapping of type (M) (see [1] for an example), however, the following proposition holds.

PROPOSITION 3 [23]. *Let* X *be a Banach space and suppose* $T: X \longrightarrow X^*$ *is a mapping of type* (M). *Suppose that* $S: X \longrightarrow X^*$ *is weakly sequentially continuous (i.e.* $u_n \rightharpoonup u$ *implies* $Su_n \longrightarrow Su$*) and that the functional* $f(u) = (Su,u)$ *is weakly sequentially lower semi-continuous on* X *(i.e.* $u_n \rightharpoonup u$ *implies* $f(u) \leq \lim\inf f(u_n)$*). Then the mapping* $T+S$ *is of type* (M).

PROOF. Let $\{u_n\}$ be a sequence in X such that

$$u_n \rightharpoonup u, \quad (T+S)u_n \longrightarrow w$$

and

$$\lim\sup((T+S)u_n , u_n) \leq (w,u).$$

Since S is weakly sequentially continuous we have that $Su_n \rightharpoonup Su$ and hence $Tu_n \longrightarrow w-Su$. Using now, the weak sequential lower semi-continuity of the functional $f(u) = (Su,u)$ and the assumption that

$$\lim\sup((T+S)u_n , u_n) \leq (w,u)$$

we obtain that

$$\lim\sup(Tu_n , u_n) \leq (w-Su , u).$$

Hence $Tu = w - Su$ (since T is of type (M)) and so $w = Tu+Su$. This shows that $T+S$ satisfies (M_1). Also it is immediate from the facts that T is of type (M) and S is weakly sequentially continuous that $T+S$ satisfies (M_2). Hence $T+S$ is of type (M) and the proof of the proposition is complete. Q.E.D.

COROLLARY 1. *Let* X *be a Banach space and suppose that* $T: X \to X^*$ *is a mapping of type* (M). *Suppose that* $S: X \to X^*$ *is a monotone, weakly sequentially continuous mapping. Then the mapping* $T+S$ *is of type* (M).

PROOF. In view of Proposition 3, it suffices to verify that the functional $f(u) = (Su,u)$ is weakly sequentially lower semi-continuous on X. So, let $u_n \to u$ (weakly) in X. Now since S is monotone and weakly sequentially continuous we have $Su_n \to Su$ and

$$0 \leqslant (Su_n - Su\,,\,u_n - u) = (Su_n\,,\,u_n) - (Su_n\,,\,u) - (Su\,,\,u_n) + (Su\,,\,u).$$

This gives on taking limits as $n \to \infty$ that

$$(Su\,,\,u) \leqslant \liminf\,(Su_n\,,\,u_n),$$

proving thereby that $f(u) = (Su\,,\,u)$ is weakly sequentially lower semi-continuous on X. Hence the corollary. Q.E.D.

COROLLARY 2. *Let* X *be a Banach space and* $T: X \to X^*$ *a mapping of type* (M). *Suppose that* $S: X \to X^*$ *is completely continuous (i.e. for any sequence* $\{u_n\}$ *in* X *such that* $u_n \to u$ *(weakly) we have* $Su_n \to Su$). *Then* $T+S$ *is a mapping of type* (M).

PROOF. Let $\{u_n\}$ be a sequence in X such that $u_n \to u$ (weakly) in X. Then $Su_n \to Su$ and so $(Su_n\,,\,u_n) \to (Su\,,\,u)$, and hence the functional $f(u) = (Su\,,\,u)$ is weakly sequentially lower semi-continuous in X. The corollary then immediately follows from Proposition 3. Q.E.D.

COROLLARY 3. *Let* X *be a Banach space and* $T: X \longrightarrow X^*$ *is a mapping of type* (M). *Suppose that* $S: X \longrightarrow X^*$ *is a monotone weakly sequentially continuous mapping and* $L: X \longrightarrow X^*$ *a completely continuous mapping. Then the mapping* T+S+L *is a mapping of type* (M).

PROOF. It follows from the proofs of Corollaries 2 and 3 that the functional $f(u) = (Su+Lu, u)$ is weakly sequentially lower semi-continuous on X. The corollary is then immediate from Proposition 3. Q.E.D.

REMARK 1. We may note that an everywhere defined single-valued monotone linear mapping $S: X \longrightarrow X^*$ is bounded and hence is weakly continuous. Hence the sum of a mapping of type (M) and an everywhere defined single-valued monotone linear mapping is a mapping of type (M) in view of Corollary 3.

Petryshyn-Fitzpatrick gave an example (in [23]) of a compact mapping in the Hilbert space ℓ_2 of square summable sequences which is not of type (M). This implies that the sum of a mapping of type (M) and a compact mapping may not be of type (M). However, the following proposition is true, but we first need a definition to state the proposition.

DEFINITION 2. A mapping $T: X \longrightarrow X^*$ is said to satisfy condition (S+) if for any sequence $\{u_j\}$ in X such that $u_j \longrightarrow u$ (weakly) in X and $\limsup (Tu_j, u_j - u) \leq 0$ then $u_j \longrightarrow u$ (strongly) in X and

$Tu_j \longrightarrow Tu$ (weakly) in X^*. Observe that a mapping satisfying condition (S+) is of type (M) while the converse is not true in general (for example, the mapping $-I$, the negative of the identity mapping, in a Hilbert space is of type (M) but does not satisfy condition (S+)).

PROPOSITION 4. *Let* X *be a Banach space,* $T: X \longrightarrow X^*$ *a bounded mapping satisfying condition* (S+) *and* $C: X \longrightarrow X^*$ *a compact mapping. Then the mapping* $T+C: X \longrightarrow X^*$ *is of type* (M).

PROOF. Let $\{u_n\}$ be a sequence in X such that $u_n \longrightarrow u$ (weakly) in X, $Tu_n + Cu_n \longrightarrow w$ (weakly) in X^* and

$$\limsup (Tu_n+Cu_n, u_n) \leq (w,u).$$

Since, now, C is compact there is a y in X^* such that $Cu_n \longrightarrow y$ in X^*. Then $Tu_n \longrightarrow w-y$ (weakly) in X^* and

$$\limsup (Tu_n, u_n-u) \leq 0$$

and so $u_n \longrightarrow u$ and $Tu_n \longrightarrow Tu$ since T satisfies condition (S+). It then follows from the continuity of C that $Cu_n \longrightarrow Cu$ and so $w = Tu + Cu$. Thus $T+C$ satisfies (M_1) in Definition 1. To verify (M_2) we observe that since T is bounded and satisfies condition (S+), T is demi-continuous and hence $T+C$ is demi-continuous from X into X^*. This immediately implies that $T+C$ satisfies (M_2) and the proof of the proposition is complete. Q.E.D.

COROLLARY 4. *Let* H *be a Hilbert space and* $C: H \longrightarrow H$ *a compact mapping. Then the mapping* $I+C: H \longrightarrow H$, *is a mapping of type* (M). (*Here* I *denotes the identity mapping in* H.)

PROOF. It suffices to observe that I is a bounded mapping satisfying condition (S+). Indeed, if $u_j \rightharpoonup u$ and

$$\limsup (u_j , u_j - u) \leq 0$$

we have

$$\|u\|^2 \leq \liminf \|u_j\|^2 \leq \limsup \|u_j\|^2 \leq \|u\|^2$$

so that $\|u_j\| \longrightarrow \|u\|$ which immediately implies that $u_j \longrightarrow u$. Hence the corollary. Q.E.D.

THEOREM 1 ([9], [23]). *Let* X *be a reflexive Banach space and let* T *be a bounded mapping of type* (M) *from* X *into* X^*. *Suppose that the mapping* T *is coercive, i.e.*

$$\lim_{\|u\| \to \infty} \frac{(Tu , u)}{\|u\|} = \infty .$$

Then $R(T) = X^*$.

PROOF. Since for any w in X^*, the mapping T_w defined by

$$T_w(x) = Tx - w$$

is a bounded coercive mapping of type (M) it suffices to prove that $0 \in R(T)$. Let

$$\Lambda = \{F \mid F \text{ a finite dimensional subset of } X \text{ and } 0 \in F\}$$

and C_F denote the convex hull of F. Since the mapping T is coercive and is continuous from finite dimensional subspaces of X to X^* endowed with weak topology it follows from Proposition 7.3 of Browder [5] that there exists a constant $R > 0$ and an element $u_F \in C_F$ for each $F \in \Lambda$

such that $\|u_F\| \leq R$ and $(Tu_F, u-u_F) \geq 0$. Let, now,

$$M = \inf\{(Tu_F, u_F) \mid F \in \Lambda\}.$$

Clearly $-\infty < M$ since the subset $\{u_F \mid F \in \Lambda\}$ is a bounded subset of X and T is a bounded mapping.

For $F_o \in \Lambda$, set

$$V_{F_o} = \{u_F \mid F \in \Lambda, \; F \supset F_o\},$$

so that V_{F_o} is contained in the ball of radius R in X, since $\|u_F\| \leq R$ for every $F \in \Lambda$. Accordingly, the family $\{\bar{V}_{F_o} \mid F_o \in \Lambda\}$ (where $\bar{V}_{F_o}$ denotes the weak closure of V_{F_o} in X) is a family of weakly compact subsets of X, since X is reflexive. So

$$\bigcap_{F_o \in \Lambda} \bar{V}_{F_o} \neq \emptyset,$$

since the family $\{\bar{V}_{F_o} \mid F_o \in \Lambda\}$ obviously has finite intersection property. Let now,

$$u_o \in \bigcap_{F_o \subset \Lambda} \bar{V}_{F_o}.$$

We assert that $Tu_o = 0$. Suppose, on the contrary, that $Tu_o \neq 0$ and let $x \in X$ be such that $(Tu_o, x) < M$. Let $F_1 \in \Lambda$ be such that $x \in F_1$ and $u_o \in F_1$. Since $u_o \in \bar{V}_{F_1}$ it follows by Proposition 7.2 of [5] that there exists an infinite sequence $\{F_i\}$, $i = 2,3,\ldots$, $F_i \in \Lambda$, $F_i \supset F_1$ for each i such that $u_{F_i} \longrightarrow u_o$ (weakly). Since T is bounded, we may assume that there is an element v_o in X^* such that $Tu_{F_i} \longrightarrow v_o$ (weakly) in X^*. It then follows easily from the relation

$$(Tu_F , u_F - u) \leq 0$$

for $F \in \Lambda$ that

$$\limsup (Tu_{F_i}, u_{F_i}) \leq (v_o , u)$$

for every $u \in C_{F_1}$. Taking $u = u_o$ this gives that $Tu_o = v_o$. Again, taking $u = x \in F_1 \subset C_{F_1}$ we get that

$$M \leq \limsup (Tu_{F_i}, u_{F_i}) \leq (v_o , x) = (Tu_o , x) < M$$

which is a contradiction. Thus $Tu_o = 0$ and the proof of the theorem is complete. Q.E.D.

§2. HAMMERSTEIN EQUATIONS IN BANACH SPACES

THEOREM 2. *Let* X *be a reflexive Banach space,* A *a monotone linear mapping from domain* $D(A)$ *in* X *into* X^* *such that the inverse mapping* A^{-1} *is defined on all of* X^* *and* A^{-1} *is single-valued. Let* N *be a bounded coercive mapping of type* (M) *from* X^* *into* X. *Then the range,* $R(I+AN)$, *of the mapping* $I+AN$ *is all of* X^*. *(Here* I *denotes the identity mapping on* X^*.*)*

PROOF. Since A is monotone, the inverse mapping $A^{-1}: X^* \longrightarrow X$ is a monotone linear mapping and hence is bounded. It then follows from Corollary 3 that the mapping $A^{-1}+N$ is a bounded mapping of type (M). Moreover,

$$\lim_{\|w\|\to\infty} \frac{(w\ ,\ A^{-1}w+Nw)}{\|w\|} \geqslant \lim_{\|w\|\to\infty} \frac{(w\ ,\ Nw)}{\|w\|} = \infty$$

so that $A^{-1}+N$ is coercive. Hence we have from Theorem 1 that

$$R(A^{-1} + N) = X.$$

This along with the fact that $R(A) = X^*$ gives that $R(I + AN) = X^*$. Hence the theorem. Q.E.D.

We now give some sufficient conditions under which the inverse of a monotone linear mapping with domain in X and range in X^* exists as a single-valued mapping defined from all of X^* into X.

THEOREM 3. *Let* X *be a reflexive Banach space and* A *be a monotone linear mapping with domain* $D(A)$ *in* X *into* X^*. *Then the inverse mapping* A^{-1} *exists as a single-valued mapping defined from all of* X^* *into* X *if any one of the following two equivalent conditions hold:*

(i) A *is a densely defined closed linear mapping such that the adjoint mapping* A^* *from its domain* $D(A^*)$ *in* X *to* X^* *is monotone and there is a constant* $\alpha > 0$ *such that* $\|Ax\| \geqslant \alpha\|x\|$ *for every* $x \in D(A)$.

(ii) A *is a densely defined closed linear mapping such that the adjoint mapping* A^* *from its domain* $D(A^*)$ *in* X *to* X^* *is monotone and there is a constant* $\alpha > 0$ *such that* $\|A^*x\| \geqslant \alpha\|x\|$ *for every* $x \in D(A^*)$.

PROOF. We first show that condition (i) implies that A^{-1} exists as a single-valued mapping defined on all of X^* into X and condition (ii) holds. In fact condition (i) implies by a result of Brézis [2, 3] that the mapping A is linear maximal monotone. This together with

$$\|Ax\| \geq \alpha\|x\|$$

for $x \in D(A)$ implies that $R(A) = X^*$ (see [5], [25]). Hence A^{-1} exists as a single-valued mapping (since A is one-to-one) defined from all of X^* into X. Now, since A is closed (so that $A = A^{**}$), $R(A) = X^*$ and $\|Ay\| \geq \alpha\|y\|$ for $y \in D(A)$, we have for $x \in D(A^*)$ that

$$\|A^*x\| = \sup_{y\in X} \frac{(A^*x\,,\,y)}{\|y\|} \geq \sup_{y\in D(A)} \frac{(A^*x\,,\,y)}{\|y\|}$$

$$= \sup_{y\in D(A)} \frac{(Ay\,,\,x)}{\|y\|} \geq \alpha \sup_{y\in D(A)} \frac{(Ay\,,\,x)}{\|Ay\|} = \alpha\|x\|.$$

Hence condition (ii) holds whenever (i) holds. Conversely, let A satisfy condition (ii) then we see that A^* satisfies condition (i) with A replaced by A^* and so it follows from above that

$$\|Ax\| = \|(A^*)^*x\| \geq \alpha\|x\| \quad \text{for} \quad x \in D(A).$$

Hence condition (i) holds whenever (ii) holds. This completes the proof of the theorem. Q.E.D.

REMARK 2. We may remark that the condition $\|Ax\| \geq \alpha\|x\|$ is interesting only for unbounded closed linear mappings because when $D(A) = X$ and A is monotone, the condition $\|Ax\| \geq \alpha\|x\|$ for all x implies that A is a homeomorphism of X onto X^*. Moreover, if A satisfies a slightly stronger hypothesis $(Ax\,,\,x) \geq \alpha\|x\|^2$ and $D(A) = X$ then X can be endowed with an equivalent Hilbert space norm. Our results accordingly generalize some of the results of [23].

DEFINITION 3. A linear mapping A with domain $D(A)$ in X to X^* is said to be *symmetric* if $(Ax, y) = (Ay, x)$ for all $x, y \in D(A)$. A mapping $N: X^* \longrightarrow X$ is called a *weak*-gradient* if there exists a functional $f: X^* \longrightarrow \mathbb{R}$ such that

$$(y^*, Nx^*) = \lim_{t \to 0} \frac{f(x^*+ty^*)-f(x^*)}{t}$$

for x^*, y^* in X^*. In particular, if X is reflexive the notion of weak*-gradient coincides with that of weak-gradient. We remark that if X is not reflexive and a given $f: X^* \longrightarrow \mathbb{R}$ is weakly-Gateaux differentiable then its weak-gradient is a mapping from X^* to X^{**} and not a mapping from X^* to X. Hence a weak-gradient is not a weak*-gradient in general.

THEOREM 4. *Let X be a real Banach space such that the unit ball in the dual Banach space X^* is weak*-sequentially compact. (In particular, X is either separable or reflexive.) Let $A: D(A) \subset X \longrightarrow X^*$ is a linear maximal monotone, symmetric densely defined mapping. Let $N: X^* \longrightarrow X$ be the weak*-gradient of a functional $f: X^* \longrightarrow \mathbb{R}$ which is assumed to be weak* lower semi-continuous and is such that*

$$f(x^*) \geqslant w(\|x^*\|)$$

for all x^* in X^*, where the real-valued function $w(r)$ is bounded below for $r > 0$ and $w(\|x^*\|) \longrightarrow \infty$ as $\|x^*\| \longrightarrow \infty$. Then there exists at leat one x^* in X^* such that $x^*+ANx^* = y^*$ for each given $y^* \in X^*$.

In the proof ot this theorem we need the following lemma on a splitting of the unbounded linear mapping A. This splitting is obtained following a similar splitting for bounded A obtained earlier by Browder-Gupta [7].

LEMMA 1. *Let* $A: D(A) \subset X \longrightarrow X^*$ *be a linear maximal monotone symmetric densely defined mapping. Then there exists a Hilbert space* H, *a linear mapping* $S: D(A) \longrightarrow H$ *such that* $A = S^*S$ *where* $S^*: D(S^*) \subset H \longrightarrow X^*$ *is the adjoint mapping of* S. *Further* S^* *is injective.*

PROOF. Consider in $D(A)$ the bilinear form b defined by

$$b[x,y] = (Ax , y)$$

for $x, y \in D(A)$. Now, since A is monotone we have

$$b[x , x] = (Ax , x) \geqslant 0$$

for $x \in D(A)$ and consequently one has Cauchy-Schwarz inequality

$$|b[x , y]| \leqslant (b[x , x])^{\frac{1}{2}} (b[y , y])^{\frac{1}{2}}. \tag{3}$$

Next we define the set $N = \{x \in D(A) \mid b[x , x] = 0\}$ which in view of (3) is also given by

$$N = \{x \in D(A) \mid b[x , y] = 0 \quad \text{for all} \quad y \in D(A)\}.$$

Thus N is a closed linear subspace of $D(A)$. So $D(A)/N$ is a prehilbertian space H_o with the inner produt $[\,,\,]$, defined as follows: Let $\Pi: D(A) \longrightarrow H_o$ be the canonical mapping and let $\bar{x}, \bar{y} \in H_o$. Then $[\bar{x} , \bar{y}] = b[x , y] = (Ax , y)$ where $x \in \Pi^{-1}(\bar{x})$, $y \in \Pi^{-1}(\bar{y})$. Now let us denote by H the completion of H_o with respect to this inner product.

Let us finally denote by S the composition of Π with the imbedding mapping of H_o into H and let us also use the notation $[\ ,\ ]$ for the inner product in H.

So the mapping $S: D(A) \longrightarrow H$ is densely defined in X and has a dense range by construction. It then follows that the adjoint mapping $S^*: D(S^*) \longrightarrow X^*$ is injective. Now, we claim that $R(S) \subset D(S^*)$, which implies in particular that S^* is densely defined in H. To prove the claimed inclusion let $S(x) \in R(S)$, $x \in D(A)$. The linear functional $\phi: D(A) \longrightarrow \mathbb{R}$ defined by $\phi(y) = [Sx, Sy]$ is bounded because

$$|[Sx , Sy]| = |(Ax , y)| \leqslant \|Ax\| \, \|y\|.$$

So ϕ can be extended to a bounded linear functional on X, which proves that $S(x) \in D(X^*)$ and

$$[Sx , Sy] = (S^*Sx , y) = (\phi , y)$$

for all y in D(A). On the other hand, $[Sx , Sy] = (Ax , y)$ by definition. It then follows from the density of D(A) in X that $S^*S = A$. This proves the lemma. Q.E.D.

Proof of Theorem 4. We first observe that we may assume $y^* = 0$, without any loss of generality. For if $y^* \neq 0$ we replace N by the mapping N_1 defined by $N_1(x^*) = N(x^*+y^*)$ and solve the equation

$$x^* + AN_1(x^*) = 0.$$

The mapping N_1 satisfies the same conditions as the mapping N. Let, now, H be the Hilbert space determined by Lemma 1 and let us denotes the norm in H by $|\cdot|$. Define a functional $\Phi: D(\Phi) \longrightarrow \mathbb{R}$, $D(\Phi) \subset H$

by

$$\Phi(u) = \frac{1}{2}|u|^2 + f(S^*u) \tag{4}$$

where S^* is the adjoint mapping of S defined in Lemma 1. So $D(\Phi) = D(S^*)$. In view of the hypothesis on f we obtain

$$\Phi(u) \geqslant \frac{1}{2}|u|^2 + w(\|S^*u\|).$$

Now, let $M = \inf\{w(r) \mid r > 0\}$. Then, $\Phi(u) \geqslant \frac{1}{2}|u|^2 + M$ for all $u \in D(\Phi)$ and so the functional Φ is bounded below in $D(\Phi)$. Let

$$d = \inf\{\Phi(u) \mid u \in D(\Phi)\}$$

and let $\{u_n\}$ be a minimizing sequence for Φ. It is clear that for all n, sufficiently large

$$d + 1 \geqslant \Phi(u_n) \geqslant \frac{1}{2}|u_n|^2 + w(\|S^*u_n\|) \geqslant M. \tag{5}$$

It then follows from (5) that the sequences $\{u_n\}$ and $\{S^*u_n\}$ are bounded. In view of the weak compactness of the unit ball in H and the weak* sequential compactness of the unit ball in X^* we may assume (by going to a subsequence, if necessary) that there exist elements u_o in H and y^* in X^* such that $u_n \longrightarrow u_o$ (weakly) in H and $S^*u_n \longrightarrow y^*$ (weak*) in X^*. Now we claim that $u_o \in D(S^*)$ which can be proved by showing that the linear functional $\ell: D(A) \longrightarrow \mathbb{R}$ defined by

$$\ell(x) = [u_o , Sx]$$

is bounded. That is so because

$$[u_o , Sx] = \lim_{n\to\infty}[u_n , Sx] = \lim_{n\to\infty}(S^*u_n , x) = (y^* , x)$$

and

$$|[u_o , Sx]| \leqslant \|y^*\| \, \|x\|.$$

It then follows from $[u_o , Sx] = (y^* , x)$ that $S^*u_o = y^*$. Next we prove that $\Phi(u)$ assumes its infimum at u_o, i.e. $\Phi(u_o) = d$. Indeed, since the norm in H is weakly lower semi-continuous and f is weak* lower semi-continuous we obtain that

$$\frac{1}{2}|u_o|^2 + f(S^*u_o) \leqslant \lim\inf\{\frac{1}{2}|u_n|^2 + f(S^*u_n)\} = \lim \Phi(u_n) = d.$$

Now, since Φ has a minimum at u_o, it follows that for all $u \in D(\Phi)$ we have

$$\lim_{t\to 0} \frac{\Phi(u_o+tu) - \Phi(u_o)}{t} = 0$$

whenever such a limit exists. It is, however, easy to see that this limit exists for all $u \in D(\Phi)$ and

$$\lim_{t\to 0} \frac{\Phi(u_o+tu) - \Phi(u_o)}{t} = [u_o , u] + (S^*u , N(S^*u_o)).$$

So

$$(6) \qquad [u_o , u] + (S^*u , N(S^*u_o)) = 0$$

for all $u \in D(\Phi)$. We now claim that $N(S^*u_o) \in D(A)$ and, moreover,

$AN(S^*u_c) = -S^*u_o$. This we can do if we prove that

(7) $$(Ax+S^*u_o\,,\,x-NS^*u_o) \geqslant 0$$

for all $x \in D(A)$, in view of the maximal monotonicity of A. Using (6) first with $u = u_o$ and then with $u = Sx$ we obtain

$$-(S^*u_o\,,\,NS^*u_o) = [u_o\,,\,u_o]$$

$$-(Ax\,,\,NS^*u_o) = -(S^*Sx\,,\,NS^*u_o) = [u_o\,,\,Sx].$$

This reduces (7) to the inequality

(8) $$(Ax\,,\,x) + [u_o\,,\,Sx] + (S^*u_o\,,\,x) + [u_o\,,\,u_o] \geqslant 0.$$

Now, let $\{x_n\}$ be a sequence in $D(A)$ such that $Sx_n \longrightarrow u_o$ in H (which can be done because $R(S)$ is dense in H). It is then easy to see that (8) is obtained from the inequality

$$(Ax+Ax_n\,,\,x+x_n) = |Sx+Sx_n|^2 \geqslant 0$$

by taking the limit as $n \longrightarrow \infty$. Finally, from the relation

$$S^*u_o + AN(S^*u_o) = 0$$

it follows that $x^* = S^*u_o$ is a solution of $x^*+ANx^* = 0$. This completes the proof of the theorem. Q.E.D.

REMARK 3. We may remark that Theorem 4, above is similar to Theorem 4 of [29]. The difference is that Theorem 4 of [29] is stated assuming very technical conditions both on the Banach space X and the linear mapping A and even then Theorem 4 of [29] gives only a weak or generalized solution of $x^* + ANx^* = y^*$; while we have obtained an actual solution under minimal and technically simple conditions both on the Banach space X and the mappings involved.

REMARK 4. If N is a Nemytskii mapping from $L_2(\Omega)$ into $L_2(\Omega)$ (where Ω is a measurable subset of $\mathbb{R}^n$ of finite Lebesgue measure) and N is defined by the function $f: \Omega \times \mathbb{R} \longrightarrow \mathbb{R}$ satisfying Caratheodory's conditions, then there must exist constants C_1, C_2 such that

$$|f(x,t)| \leqslant C_1 + C_2|t|$$

for $x \in \Omega$ and $t \in \mathbb{R}$. It is then well-known that N is the gradient of the functional ψ on $L_2(\Omega)$ defined by

$$\psi(u) = \int_\Omega F(x,u(x))\,dx$$

where

$$F(x,t) = \int_0^t f(x,s)\,ds$$

and $F(x,t) \geqslant -\frac{1}{2}C_2|t|^2 - C_3$. This shows that ψ satisfies an estimate of the form

$$\psi(u) \geqslant -\frac{1}{2}C_2\|u\|^2 - C_3$$

for $u \in L_2(\Omega)$.

REMARK 5. The estimate $f(x^*) \geqslant w(\|x^*\|)$ on the potential f of the Nemytskii mapping N in Theorem 4 holds, if for example f satisfies an estimate of the form

$$f(x^*) \geqslant a_1\|x^*\|^2 + a_2\|x^*\|^\theta + a_3$$

where $a_1 > 0$, $0 < \theta < 2$ and a_2, a_3 are arbitrary constants. In view of Remark 4, such an estimate is natural on the potential f of a Nemytskii mapping. This also points up the fact that the equation

$$x^* + ANx^* = y^*$$

can be solved only for a certain class of N when A is unbounded and this is only natural as the conditions on A and N have to somehow balance out the problem. (See e.g., a remark due to Browder on p. 428 of [6].)

THEOREM 5. *Let* X *be a reflexive Banach space,* A *a maximal monotone mapping from* X^* *into* 2^X *such that the mapping* $N^{-1} + A$ *from* X *into* 2^{X^*} *is maximal monotone. Suppose* k(r) *and* c(r) *are real-valued functions of* r *in* $\mathbb{R}^+$ *such that* $k(r)+c(r) \longrightarrow +\infty$ *as* $r \longrightarrow \infty$ *and that the following two inequalities hold:*

(i) $(w,u) \geqslant k(\|u\|)\|u\|$ *for* $u \in D(A)$ *and* $w \in Au$.

(ii) $(w,u) \geqslant c(\|u\|)\|u\|$ *for* $u \in X$, $w \in N^{-1}u$.

Then the range, R(I+AN), *of the mapping* I + AN *is all of* X^*.

PROOF. We first note that the solvability of the equation

$$w \in N^{-1}u + Au$$

for a given w in X^* implies the solvability of the equation

$$w \in x^* + ANx^*$$

with $x^* \in N^{-1}u$. So it suffices to show that the maximal monotone mapping $N^{-1} + A$ from X into 2^{X^*} is surjective which will be the case if $N^{-1} + A$ is coercive. Now, for $u \in X$ and $w \in N^{-1}u + Au$, so that $w = w_1 + w_2$ with $w_1 \in N^{-1}u$ and $w_2 \in Au$, we have

$$(w , u) = (w_1 , u) + (w_2 , u) \geqslant (c(\|u\|) + k(\|u\|))\|u\| .$$

This shows that $N^{-1} + A$ is coercive and hence $R(N^{-1}+A) = X^*$ or in other words, the equation $w \in N^{-1}u + Au$ is solvable for any given $w \in X^*$. This completes the proof of the Theorem. Q.E.D.

REMARK 6. We may note that by a result of Browder [5], Rockafellar [25], the mapping $N^{-1} + A$ is maximal monotone if, e.g., either $\operatorname{Int} D(A) \cap R(N)$ is non-empty or $\operatorname{Int} R(N) \cap D(A)$ is non-empty. For further conditions which ensure the maximal monotonicity of the mapping $N^{-1}+A$ we refer the reader to [4].

REMARK 7. Theorem 5 allows us to study nonlinear integral equations of Hammerstein type when N may not be defined everywhere. Such equations do arise in applications.

REMARK 8. Theorem 5 generalizes Theorem 3 of Browder-de Figueiredo-Gupta [8].

§3. HAMMERSTEIN EQUATIONS IN HILBERT SPACES

We begin this section by giving an example of an unbounded linear integral mapping in the Hilbert space $L_2[0,1]$ of square integrable functions on $[0,1]$.

EXAMPLE 1. Let $L_2[0,1]$ denote the Hilbert space of square integrable real-valued functions on $[0,1]$. Let $\{\phi_n\}$ be a sequence of orthonormal functions in $L_2[0,1]$ with pairwise disjoint supports, that is, for every $n, m, n \neq m$,

$$\{x \in [0,1] \mid \phi_n(x) \neq 0\} \cap \{x \in [0,1] \mid \phi_m(x) \neq 0\} = \emptyset.$$

(For an example of such a sequence see Example A, §9, Chapter 9 of Zaanen [30].) Let H be the closed subspace of $L_2[0,1]$ spanned by the sequence $\{\phi_n\}$. Let $\{\lambda_n\}$ be any given sequence of real numbers and let $K(x,y)$ be defined on $[0,1] \times [0,1]$ by

$$K(x,y) = \sum_{n=1}^{\infty} \lambda_n \phi_n(x) \phi_n(y).$$

Observe that for each $(x,y) \in [0,1] \times [0,1]$ at most one term of the series

$$\sum_{n=1}^{\infty} \lambda_n \phi_n(x) \phi_n(y)$$

is different from zero. Let A be the linear integral operator on $L_2[0,1]$ having $K(x,y)$ as its kernel function. Thus for $f \in L_2[0,1]$ we have

$$(Af)(x) = \int_0^1 K(x,y) f(y)\,dy = \sum_{n=1}^{\infty} \lambda_n (f,\phi_n) \phi_n(x),$$

where (f, ϕ_n) denotes the inner product in $L_2[0,1]$. We note that the domain $D(A)$ of A is given by

$$D(A) = \{f \in L_2[0,1] \mid \sum_{n=1}^{\infty} |\lambda_n|^2 |(f, \phi_n)|^2 < +\infty\}.$$

Clearly, $\phi_n \in D(A)$ for every $n = 1,2,\ldots$ and the orthogonal complement $H^{\perp}$ of H in $L_2[0,1]$ is contained in $D(A)$. Thus $D(A)$ is dense in $L_2[0,1]$ and the range of A is contained in the subspace H of $L_2[0,1]$. We may remark that the operator A is not bounded, in general, since we can have that $|\lambda_n| \longrightarrow \infty$ as $n \longrightarrow \infty$. However, we have the following proposition.

PROPOSITION 5. *The linear integral operator* $A: D(A) \longrightarrow L_2[0,1]$ *defined above, is closed.*

PROOF. Let $\{f_n\}$ be a sequence in $D(A)$ and $f, g \in L_2[0,1]$ be such that $f_n \longrightarrow f$, $Af_n \longrightarrow g$. Since for every n, $Af_n \in H$ we have that $g \in H$. Moreover, since $\{\phi_n\}$ is a complete orthonormal basis for H we have

$$g = \sum_{i=1}^{\infty} (g, \phi_i)\phi_i \quad \text{and} \quad \|g\|^2 = \sum_{i=1}^{\infty} |(g, \phi_i)|^2.$$

Now, for every n, we have

$$Af_n = \sum_{i=1}^{\infty} \lambda_i (f_n, \phi_i)\phi_i$$

and then $(Af_n, \phi_j) = \lambda_j (f_n, \phi_j)$ for every n, $j = 1,2,\ldots$. Taking limits as $n \longrightarrow \infty$ we get $(g, \phi_j) = \lambda_j (f, \phi_j)$ for every j. Hence

$$\sum_{j=1}^{\infty} |\lambda_j|^2 |(f , \phi_j)|^2 = \sum_{j=1}^{\infty} |(g , \phi_j)|^2 = \|g\|^2.$$

So $f \in D(A)$ and

$$Af = \sum_{j=1}^{\infty} \lambda_j (f , \phi_j)\phi_j = \sum_{j=1}^{\infty} (g , \phi_j)\phi_j = g.$$

Hence the proposition. Q.E.D.

PROPOSITION 6. *The linear integral operator* $A: D(A) \longrightarrow L_2[0,1]$ *is self-adjoint.*

PROOF. Since $f, g \in D(A)$ we see easily that $(Af,g) = (f,Ag)$ we have that A is symmetric, i.e., $D(A) \subset D(A^*)$ and A^* restricted to $D(A)$ coincides with A. So it suffices to prove that $D(A^*) \subset D(A)$. Indeed, let $f \in D(A^*)$. Then there exists a constant C (depending on f) such that $|(Ag , f)| \leqslant C\|g\|$ for every $g \in D(A)$. Taking for each $n = 1,2,\ldots,$

$$g_n = \sum_{i=1}^{n} \lambda_i (f , \phi_i)\phi_i \in D(A)$$

we get that

$$\sum_{i=1}^{n} |\lambda_i|^2 |(f , \phi_i)|^2 = |(Ag_n , f)| \leqslant C\|g_n\| = C\{\sum_{i=1}^{n} |\lambda_i|^2 |(f,\phi_i)|^2\}^{\frac{1}{2}}.$$

This implies that, for every $n = 1,2,\ldots,$

$$\sum_{i=1}^{n} |\lambda_i|^2 |(f , \phi_i)|^2 \leqslant C^2 .$$

and hence

$$\sum_{i=1}^{\infty} |\lambda_i|^2 |(f, \phi_i)|^2 \leqslant c^2.$$

So $f \in D(A)$ and hence $D(A) \supset D(A^*)$. This completes the proof of the proposition. Q.E.D.

REMARK 9. Note that Propostiion 5 is a corollary of Proposition 6 since every self-adjoint densely defined linear mapping is closed.

REMARK 10. If we assume that $\lambda_n \geqslant 0$ for all n, then the linear integral operator A affords an example of an unbounded linear maximal monotone operator in $L_2[0,1]$. This follows from the fact that A is self-adjoint using a result of Brézis [2].

THEOREM 6. *Let* H *be a Hilbert space and* A *a (closed) linear maximal monotone mapping from domain* D(A) *in* H *to* H. *Let* N *be a bounded mapping of type* (M) *from* H *into* H *such that*

$$\liminf_{\|u\| \to \infty} (N(u+v), u) > 0$$

for every $v \in H$. *Then the equation* $u + ANu = v$ *is solvable for each given* $v \in H$, *that is, the mapping* $I + AN$ *is surjective.*

PROOF. As in the proof of Theorem 4, it suffices to show that the equation $u + ANu = 0$ is solvable in H. Now, to prove the solvability of the equation $u + ANu = 0$ we first observe that there exists an $r > 0$ such that $(Nu, u) > 0$ for $\|u\| \geqslant r$. Next, since A is a closed linear maximal monotone mapping it follows, from a result of Phillips

[24], that for each $n = 1,2,...$ $(A + \frac{1}{n}I)^{-1}$ exists as a bounded linear mapping from H into H. It then follows from Corollary 1 that $(A + \frac{1}{n}I)^{-1} + N$ is a bounded mapping of type (M) and

$$((A + \frac{1}{n}I)^{-1}u + Nu\ , u) > 0$$

for all $\|u\| \geqslant r$. Hence, for each n, there exists a $u_n \in H$ with

$$\|u_n\| < r \quad \text{and} \quad (A + \frac{1}{n}I)^{-1}u_n + Nu_n = 0,$$

by Theorem 1. Since the mapping N is bounded we may assume (by going to a subsequence, if necessary) that there exist elements u, v in H such that $u_n \longrightarrow u$ (weakly) and $Nu_n \longrightarrow v$ (weakly) in H. We now assert that $v \in D(A)$ and $Av = -u$. To prove this we first observe that the graph of A is weakly closed in $H \times H$, since A is linear maximal monotone. It then follows from

$$u_n + ANu_n + \frac{1}{n}Nu_n = 0,$$

$u_n \longrightarrow u$ and $Nu_n \longrightarrow v$ that $ANu_n \longrightarrow -u$ and this gives that $v \in D(A)$ with $Av = -u$. We now assert that $Nu = v$ which would imply that $ANu = -u$ or $u + ANu = 0$ and the proof of the theorem would be complete. Now to prove the assertion, we have, using the monotonicity of A, that

$$0 \leqslant (Av-ANu_n\ , v-Nu_n) \leqslant (Av\ , v-Nu_n) + (u_n\ , v-Nu_n) + \frac{1}{n}(Nu_n\ , v-Nu_n).$$

This gives that

$$(u_n\ , Nu_n) \leqslant (Av\ , v-Nu_n) + (u_n\ , v) + \frac{1}{n}(Nu_n\ , v-Nu_n)$$

and so $\limsup\,(u_n\,,Nu_n) \leqslant (u\,,v)$. Hence $v = Nu$ since N is of type (M). This proves the assertion and the proof of the theorem is complete. Q.E.D.

REMARK 11. Note that the solution of the equation $u + ANu = v$ is unique if N satisfies the additional condition

$$(u-v\,,Nu-Nv) \leqslant 0$$

implies that $Nu = Nv$. Indeed, if $u_1, u_2 \in H$ are such that

$$u_1 + ANu_1 = v = u_2 + ANu_2$$

we have

$$0 = (u_1-u_2\,,Nu_1-Nu_2) + (ANu_1-ANu_2\,,Nu_1-Nu_2) \geqslant (u_1-u_2\,,Nu_1-Nu_2).$$

So $Nu_1 = Nu_2$ and then $u_1 = u_2$ since

$$u_1 + ANu_1 = u_2 + ANu_2 \quad \text{and} \quad ANu_1 = ANu_2.$$

REMARK 12. Compare Theorem 6 with Theorem 2 where we have an additional assumption on A, viz. that A^{-1} exists as a bounded linear monotone mapping from X^* to X. This indicates a certain gap between the results for Banach and Hilbert space cases. For an analog of Theorem 6 for Banach spaces without any additional assumptions on A we refer the reader to Brézis [4], Gupta [13] and Kenmochi [17].

THEOREM 7[19]. *Let* H *be a Hilbert space and* A *a closed linear maximal monotone mapping from domain* D(A) *in* H *to* H. *Let* N *be a bounded hemi-continuous mapping from* H *into* H *such that there exists a completely continuous mapping* G *from* H *into* H *such that*

$$(Nx-Ny\ ,x-y) \geqslant (Gx-Gy\ ,x-y)$$

for $x, y \in H$. *Suppose further that*

$$\lim_{\|u\|\to\infty} \frac{(N(u+v)\ ,u)}{\|u\|} = \infty$$

for each $v \in H$. *Then the equation* $u + ANu = v$ *is solvable in* H *for each given* v *in* H.

PROOF. In view of Theorem 6, it suffices to show that N is a mapping of type (M) which follows immediately from Corollary 2 since N = (N-G) + G and N-G being hemi-continuous monotone is of type (M). This completes the proof. Q.E.D.

THEOREM 8. *Let* H *be a given Hilbert space,* A *a closed linear maximal monotone mapping from domain* D(A) *in* H *into* H, B *a compact (possibly nonlinear) mapping from* H *into* H *and* N *a bounded hemi-continuous strongly monotone mapping (i.e., there exists* $\alpha > 0$ *such that* $(Nu-Nv\ ,u-v) \geqslant \alpha\|u-v\|^2$ *for* $u, v \in H$) *from* H *into* H. *Suppose that the following conditions hold:*

(i) *there exist constants* $\beta > 0$, $c > 0$ *such that* $\|Nu\| \leqslant \beta\|u\| + c$ *for* u *in* H,

(ii) *there exist constants* $\gamma > 0$, $d > 0$ *such that* $\|Bu\| \leqslant \gamma\|u\| + d$

for u *in* H,

and

(iii) $\gamma(\beta + \beta^2/\alpha) < 1$.

Then the range $R(I+(A+B)N) = H$.

We need the following proposition in the proof of Theorem 8.

PROPOSITION 7, *Let* H *be a given Hilbert space,* A *a closed linear maximal monotone mapping from domain* $D(A)$ *in* H *into* H. *Let* N *be a bounded hemi-continuous strongly monotone mapping (i.e., there exists* $\alpha > 0$ *such that* $(Nu-Nv\,,u-v) \geqslant \alpha\|u-v\|^2$ *for* u, v *in* H). *Then the inverse mapping* $T^{-1} = (I+AN)^{-1}$ *exists as a bounded continuous mapping from* H *into* H.

PROOF. We first note from Theorem 6, the strong-monotonicity of N and Remark 11 that the mapping $T = I + AN$ is one-to-one and $R(I+AN) = H$. We shall show that T^{-1} is bounded and continuous on every ball of radius r and center $0 \in H$ in H. So let $r > 0$ be given and $f, g \in H$ be such that $\|f\| \leqslant r$, $\|g\| \leqslant r$. Then there exist u, v in H such that $u + ANu = f$ and $v + ANv = g$. Thus we have

$$(u-v\,,Nu-Nv) + (ANu-ANv\,,Nu-Nv) = (f-g\,,Nu-Nv)$$

which implies that

$$\alpha\|u-v\|^2 \leqslant (f-g\,,Nu-Nv) \leqslant \|f-g\|\,\|Nu-Nv\|. \tag{9}$$

On the other hand, we have $0 = (u-f, Nu) + (ANu, Nu)$ and so

$$(u-f, Nu) \leqslant 0. \tag{10}$$

Hence,

$$\alpha\|u-f\|^2 \leqslant (u-f, Nu-Nf) \leqslant -(u-f, Nf) \leqslant \|u-f\| \, \|Nf\|.$$

So,

$$\alpha\|u-f\| \leqslant \|Nf\|. \tag{11}$$

Similarly, $\alpha\|v-g\| \leqslant \|Ng\|$. Using (9), (10) and (11) and the boundedness of N we see that there is a constant $M > 0$ (depending on r) such that

$$\|u-v\|^2 \leqslant \frac{2}{\alpha} M\|f-g\|.$$

This shows that T^{-1} is bounded and continuous on every ball of radius r and center $0 \in H$. Hence the proposition. Q.E.D.

Proof of Theorem 8: We first see from Proposition 7 that the inverse mapping $T^{-1} = (I+AN)^{-1}$ is a bounded continuous mapping from H into H and the mapping BNT^{-1} is compact. Now, since

$$I + (A+B)N = (I+BNT^{-1})T$$

and T is surjective, it suffices to show that $R(I+BNT^{-1}) = H$. This will follow from Theorem 1 if we show that $I+BNT^{-1}$ is coercive since we have from Corollary 4 that $I+BNT^{-1}$ is a bounded mapping of type (M). But it is easy to see from our assumptions that

$$\lim_{\|u\|\to\infty} \frac{(u+BNT^{-1}u\,,\,u)}{\|u\|} \geq \lim_{\|u\|\to\infty} \{(1-\gamma|\beta+\frac{\beta^2}{\alpha}|)\|u\| - C\} = \infty.$$

This completes the proof of the theorem. Q. E. D.

REMARK 13. Theorem 8 generalizes some results of Vainberg [27] for the case of Hammerstein equations involving indefinite linear mappings which are finite dimensional perturbations of monotone linear mappings.

The following theorem deals with the case when the negative part of an indefinite linear mapping is unbounded and non-compact.

THEOREM 9. *Let* H *be a Hilbert space,* H_1 *a closed subspace of* H *and* H_2 *be its orthogonal complement. Let* $N: H \longrightarrow H$ *be a continuous, or bounded demi-continuous, strongly monotone mapping with* $\alpha > 0$ *as the constant of strong monotonicity of* N*. Let* $A_i: D(A_i) \longrightarrow H_i\,,\, D(A_i) \subset H_i$ *be closed linear maximal monotone mappings in* H_i *for* $i = 1,2$*. Moreover, assume that there is a constant* $m > 0$ *such that* $\alpha m > 1$ *and*

$$(A_2u\,,\,u) \geq m\|u\|^2$$

for all $u \in D(A_2)$*. Then the equation*

$$u + (A_1P_1 - A_2P_2)Nu = f$$

has exactly one solution in H *for each* f *in* H*. (Here* P_i *denotes the orthogonal projection of* H *onto* H_i *for* $i = 1,2$*.)*

PROOF. As in the proof of Theorem 4, it suffices to prove that the equation

$$u + (A_1P_1 - A_2P_2)Nu = 0 \tag{12}$$

has exactly one solution in H. The solvability of equation (12) is equivalent to the solvability of the system of equations

$$u_1 + A_1P_1N(u_1 + u_2) = 0 \tag{13}$$

$$u_2 - A_2P_2N(u_1 + u_2) = 0 \tag{14}$$

where $u_i \in H_i$ for $i = 1,2$.

We first assert that for each fixed u_1 in H_1 there exists a unique u_2 in H_2 which solves equation (14). Indeed, it is easy to see that the mapping $S_{u_1}: H_2 \longrightarrow H_2$ defined by

$$S_{u_1}(u_2) = P_2N(u_1 + u_2)$$

is demi-continuous and strongly-monotone with constant α. So the mapping S_{u_1} is surjective on H_2 and has a Lipschitzian (single-valued) inverse with Lipschitz constant $\frac{1}{\alpha}$. Also since A_2 is maximal monotone and satisfies

$$(A_2u , u) \geqslant m\|u\|^2$$

for $u \in D(A_2)$ we see that A_2 is surjective on H_2 and also has a Lipschitzian (single-valued) inverse with Lipschitz constant $\frac{1}{m}$. Hence (14) is equivalent to the equation

$$u_2 = S_{u_1}^{-1} A_2^{-1} u_2. \tag{15}$$

Since $\alpha m > 1$, we see that (15) has a unique solution in H_2 by contraction mapping principle. We thus define a mapping $R: H_1 \to H_2$ which associates to each $u_1 \in H_1$ the unique element u_2 $(=Ru_1)$ in H_2 which solves (14). That is we have

$$Ru_1 - A_2 P_2 N(u_1 + Ru_1) = 0. \tag{16}$$

We now assert that the mapping $R: H_1 \longrightarrow H_2$ defined above is continuous in the norm topologies under the conditions of the theorem.

Case 1: N continuous.

By the strong monotonicity of N we have for $u_1, v_1 \in H_1$ that

$$\begin{aligned} \alpha \| Ru_1 - Rv_1 \|^2 \leq\ & (N(u_1+Ru_1) - N(v_1+Rv_1),\ Ru_1 - Rv_1) + \\ & + (N(v_1+Rv_1) - N(u_1+Rv_1),\ Ru_1 - Rv_1). \end{aligned} \tag{17}$$

Using (16), it follows that the first term on the right-hand side of (17) is equal to

$$(A_2^{-1} Ru_1 - A_2^{-1} Rv_1,\ Ru_1 - Rv_1) \leq \frac{1}{m} \| Ru_1 - Rv_1 \|^2. \tag{18}$$

Hence we get from (17) that

$$(19)\qquad (\alpha - \tfrac{1}{m})\|Ru_1 - Rv_1\| \leqslant \|N(v_1+Rv_1) - N(u_1+Rv_1)\|$$

and then the continuity of R at $v_1 \in H_1$ follows from the continuity of N at $v_1+Rv_1 \in H$.

Case 2: N bounded demi-continuous.

We first observe from (19) by taking $v_1 = 0$ that the boundedness of N implies the boundedness of the mapping R. Now, again from the strong monotonicity of N and the orthogonality of H_1 and H_2, we see that

$$(20)\qquad \|u_1-v_1\|^2 + \alpha\|Ru_1-Rv_1\|^2 \leqslant (N(u_1+Ru_1)-N(v_1+Rv_1),\ u_1-v_1) + \\ + (N(u_1+Ru_1)-N(v_1+Rv_1),\ Ru_1-Rv_1)$$

which gives using (16) and (18) that

$$(21)\qquad \alpha\|u_1-v_1\|^2 + (\alpha-\tfrac{1}{m})\|Ru_1-Rv_1\|^2 \leqslant (N(u_1+Ru_1)-N(v_1+Rv_1),\ u_1-v_1).$$

This inequality implies the continuity of R (since both N and R are bounded in this case) at $v_1 \in H_1$.

To finish the proof we have only to prove that the equation

$$(22)\qquad u_1 + A_1P_1N(u_1 + Ru_1) = 0$$

has exactly one solution in H_1. To prove this we first observe from the continuity of R and the demi-continuity of N that the mapping

$Tu_1 = P_1N(u_1+Ru_1)$ is demi-continuous from H_1 into H_1. Next we prove that T is strongly monotone. Indeed, we observe from (21) that

$$\alpha \| u_1 - v_1 \|^2 \leqslant (Tu_1 - Tv_1 \, , \, u_1 - v_1)$$

for all $u_1, v_1 \in H_1$. So the equation (22) which can be written as

$$u_1 + A_1 Tu_1 = 0$$

has at leat one solution in H_1 by Theorem 1 of Browder-de Figueiredo-Gupta [8]. We now prove that the solution of (22) is unique. Indeed, let $u_1, v_1 \in H_1$ be two solutions of (22), then from the strong monotonicity of T we have

$$0 = (u_1 - v_1 + A_1 Tu_1 - A_1 Tv_1 \, , \, Tu_1 - Tv_1) \geqslant \alpha \| u_1 - v_1 \|^2$$

which implies that $u_1 = v_1$. This completes the proof of the theorem. Q.E.D.

REMARK 14. We may note that in the proof of Theorem 9, above, the maximal monotonicity of A_2 and the condition $(A_2 u \, , \, u) \geqslant m \| u \|^2$ for $u \in D(A_2)$ were used to assert that A_2 is surjective. This fact enables us to state that one could replace the assumptions on A_2 by either one of the following weaker conditions:

(i) A_2^{-1} exists as a bounded linear mapping defined on all of H_2 with

$$\| A_2^{-1} \| \leqslant \frac{1}{m}$$

or

(ii) $A_2 \colon D(A_2) \longrightarrow H_2$ is linear maximal monotone with

$$\|A_2u\| \geqslant m\|u\|$$

for $u \in D(A_2)$.

REMARK 15. Theorem 9 is similar to Theorem 4 of Koscikii [19].

REMARK 16. An example of a linear mapping A $(= A_1P_1 - A_2P_2)$ that appears in Theorem 9 above is obtained, for example by taking $\lambda_{2i} \geqslant 0$, $\lambda_{2i-1} \leqslant 0$ with $\lambda_{2i-1} \leqslant -m$ in the example discussed in the beginning of this section.

We shall now study a nonlinear equation of Hammerstein type involving unbounded indefinite linear mapping in a Hilbert space using variational methods. We remark that our results in this case are much better than the corresponding results that we had in a Banach space in Section 1. We obtain generalizations of some of the results of Vainberg [27], Lavrentiev [21], [22], Vainberg-Lavrentiev and others. (See also de Figueiredo-Gupta [11].)

Let H be a real Hilbert space. We first remark that if $A: D(A) \longrightarrow H$, where $D(A)$ is a dense subspace of H, is a self-adjoint, monotone linear mapping then A is closed and m-accretive in the sense of Kato [16] and so has a unique m-accretive square root $A^{1/2}$ which is also self-adjoint.

Let H_1, H_2 be closed orthogonal subspaces of H such that $H = H_1 + H_2$ and let P_i denote the orthogonal projections of H onto H_i for $i = 1,2$. Let $A: D(A) \longrightarrow H$ be a densely defined self-adjoint linear mapping such that $P_i(D(A)) \subset D(A)$ for $i = 1,2$ so that

$A = AP_1 + AP_2$ and $P_iA = AP_i$ for $i = 1,2$. Further, suppose that

$$(AP_1x, x) \geqslant 0, \quad (AP_2x, x) \leqslant 0 \quad \text{for} \quad x \in D(A).$$

We then conclude from the boundedness of the projections P_i, $i = 1,2$, that AP_1, AP_2, $|A| = AP_1 - AP_2$ are densely defined self-adjoint monotone linear mappings from $D(A)$ into H. Next by a straightforward calculation we obtain:

$$|A|^{\frac{1}{2}} = (AP_1)^{\frac{1}{2}} + (-AP_2)^{\frac{1}{2}} \tag{23}$$

$$\left[(AP_1)^{\frac{1}{2}} - (-AP_2)^{\frac{1}{2}}\right] (P_1-P_2) = |A|^{\frac{1}{2}} \tag{24}$$

$$\left[(AP_1)^{\frac{1}{2}} - (-AP_2)^{\frac{1}{2}}\right] |A|^{\frac{1}{2}} = |A|^{\frac{1}{2}}\left[(AP_1)^{\frac{1}{2}} - (-AP_2)^{\frac{1}{2}}\right] = A. \tag{25}$$

With these preliminary remarks we can now state and prove our next theorem.

THEOREM 10. *Let* H, H_i, P_i *and* A *be as described above. Assume further that* $\dim H_2$ *is finite and*

$$(AP_2x, x) \leqslant -m\|x\|^2$$

for all $x \in D(A)$ *and some* $m > 0$. *Let* $f: H \longrightarrow \mathbb{R}$ *be a weakly Gateaux differentiable function which is also weakly lower semi-continuous and is such that*

$$f(x) \geqslant \frac{1}{2} a_1\|x\|^2 + a_2\|x\|^\theta + a_3$$

where $a_1 m > 1$, $a_2 \leqslant 0$, $a_3 \leqslant 0$ *and* $0 \leqslant \theta < 2$.

Then the equation $x + ANx = y$ *has at leat one solution in* H *for each given* y *in* H. *(Here* N *denotes the weak-gradient of* f.*)*

PROOF. As in the proof of Theorem 4, we may assume without any loss of generality that $y = 0$. Define in $D(|A|^{1/2})$ the following inner product

$$[x , y] = (x,y) + (|A|^{\frac{1}{2}}x , |A|^{\frac{1}{2}}y)$$

which gives the norm

$$|x| = (\|x\|^2 + \| |A|^{\frac{1}{2}}x\|^2)^{\frac{1}{2}},$$

$x, y \in D(|A|^{\frac{1}{2}})$. The linear subspace $D(|A|^{\frac{1}{2}})$ with this new norm is complete and consequently is a Hilbert space which we denote by H'. In fact, if $\{u_n\}$ is a Cauchy sequence in H' then $\{u_n\}$ and $\{|A|^{1/2}u_n\}$ are both Cauchy sequences in H. So there are u, v in H such that $u_n \longrightarrow u$ and $|A|^{1/2}u_n \longrightarrow v$ in H. Since $|A|^{1/2}$ is closed we see that $u \in D(|A|^{1/2})$ and $|A|^{1/2}u = v$. Thus $u \in D(|A|^{1/2}) = H'$ and $u_n \longrightarrow u$ in H'.

Now consider the functional $\Phi : H' \longrightarrow \mathbb{R}$ defined by

$$\Phi(u) = \frac{1}{2}\|P_1 u\|^2 - \frac{1}{2}\|P_2 u\|^2 + f(|A|^{\frac{1}{2}}u)$$

for $u \in H'$. We shall prove now that Φ is weakly lower semi-continuous on H' and $\Phi(u) \longrightarrow +\infty$ as $|u| \longrightarrow +\infty$. To do this we first note that if $u_n \longrightarrow u$ (weakly) in H' then $u_n \longrightarrow u$ (weakly) in H and $|A|^{1/2}u_n \longrightarrow |A|^{1/2}u$ (weakly) in H. In fact,

$$[u_n, v] = (u_n, v) + (|A|^{\frac{1}{2}}u_n, |A|^{\frac{1}{2}}v) = (u_n, v+|A|v)$$

for all $v \in D(|A|^{\frac{1}{2}})$. Since $D(A) \subset D(|A|^{\frac{1}{2}})$ and $R(I+|A|) = H$, by virtue of m-accretiveness of $|A|$ it follows that $u_n \longrightarrow u$ (weakly) in H. Similarly, using the relation

$$[u_n, |A|^{\frac{1}{2}}v] = (u_n, |A|^{\frac{1}{2}}v) + (|A|^{\frac{1}{2}}u_n, |A|v) = (|A|^{\frac{1}{2}}u_n, v+|A|v)$$

we prove that $|A|^{\frac{1}{2}}u_n \longrightarrow |A|^{\frac{1}{2}}u$ (weakly) in H. Thus we have

$$\|u\| \leqslant \liminf_n \|u_n\|$$

$$f(|A|^{\frac{1}{2}}u) \leqslant \liminf_n f(|A|^{\frac{1}{2}}u_n)$$

whenever $u_n \longrightarrow u$ (weakly) in H'. These two inequalities and the assumption that $\dim H_2$ is finite then imply that

$$\Phi(u) \leqslant \liminf_n \Phi(u_n)$$

whenever $u_n \longrightarrow u$ (weakly) in H'. Thus Φ is weakly lower semi-continuous on H'. Moreover, from the assumptions on f we have

$$\Phi(u) \geqslant c|u|^2 + a_2|u|^\theta + a_3$$

where

$$c = \frac{1}{2}\min(1,\ m\varepsilon,\ \varepsilon) \quad \text{and} \quad a_1 - \frac{1}{m} = 2\varepsilon,$$

which shows that $\Phi(u) \longrightarrow \infty$ as $|u| \longrightarrow \infty$ for $u \in H'$. {Indeed,

$$\Phi(u) \geqslant \frac{1}{2}\|P_1u\|^2 - \frac{1}{2}\|P_2u\|^2 + \frac{1}{2}a_1\|\,|A|^{\frac{1}{2}}u\|^2 + a_2\|\,|A|^{\frac{1}{2}}u\|^\theta + a_3$$

$$\geqslant \frac{1}{2}\|P_1u\|^2 - \frac{1}{2}\|P_2u\|^2 + \frac{1}{2m}\|\,|A|^{\frac{1}{2}}u\|^2 + \frac{\varepsilon}{2}\|\,|A|^{\frac{1}{2}}u\|^2 +$$

$$+ \frac{\varepsilon}{2}\|\,|A|^{\frac{1}{2}}u\|^2 + a_2|u|^\theta + a_3$$

$$\geqslant \frac{1}{2}\|P_1u\|^2 - \frac{1}{2}\|P_2u\|^2 + \frac{1}{2m}\|(-AP_2)^{\frac{1}{2}}u\|^2 + \frac{\varepsilon}{2}\|(-AP_2)^{\frac{1}{2}}u\|^2 +$$

$$+ \frac{\varepsilon}{2}\|\,|A|^{\frac{1}{2}}u\|^2 + a_2|u|^\theta + a_3$$

$$> c|u|^2 + a_2|u|^\theta + a_3$$

since

$$\|(-AP_2)^{\frac{1}{2}}u\|^2 = (-AP_2u\,,\,u) \geqslant m\|u\|^2 \geqslant m\|P_2u\|^2$$

for $u \in D(|A|^{\frac{1}{2}})$. } Hence, there exists a $u_o \in H'$ where Φ assumes its minimum. At this point, the weak gradient of Φ vanishes and this gives

$$(P_1 u_0 - P_2 u_0 , v) + (N|A|^{\frac{1}{2}} u_0 , |A|^{\frac{1}{2}} v) = 0$$

for all $v \in D(|A|^{\frac{1}{2}})$. Since $|A|^{\frac{1}{2}}$ is self-adjoint the above equation implies that $-N|A|^{\frac{1}{2}} u_0 \in D(|A|^{\frac{1}{2}})$ and

$$|A|^{\frac{1}{2}} N |A|^{\frac{1}{2}} u_0 = -P_1 u_0 + P_2 u_0 ,$$

that is,

$$P_1 u_0 - P_2 u_0 + |A|^{\frac{1}{2}} N |A|^{\frac{1}{2}} u_0 = 0.$$

Applying $(AP_1)^{\frac{1}{2}} - (-AP_2)^{\frac{1}{2}}$ to both sides of the above equation and using (24), (25) we then obtain that

$$|A|^{\frac{1}{2}} u_0 + AN|A|^{\frac{1}{2}} u_0 = 0.$$

Hence $x_0 = |A|^{\frac{1}{2}} u_0$ is a solution of $x + ANx = 0$. This completes the proof of the theorem. Q.E.D.

THEOREM 11. *Let* H *be a real Hilbert space and* $A: D(A) \longrightarrow H$, $D(A) \subset H$, *be a densely defined self-adjoint linear maximal monotone mapping. Let* $f: H \longrightarrow \mathbb{R}$ *be a weakly Gateaux differentiable function which is also weakly lower semi-continuous and is such that*

$$f(x) \geq \frac{1}{2} a_1 \|x\|^2 + a_2 \|x\|^{\theta} + a_3$$

where $a_1 > 0$, $a_2 \leq 0$, $a_3 \leq 0$ *and* $0 \leq \theta < 2$.

Then the equation $x + ANx = y$ *has at least one solution in* H *for each given* y *in* H. *(Here* N *denotes the weak-gradient of* f.*)*

PROOF. The proof of this theorem follows immediately from Theorem 10 if we observe that in this case $H_2 = \{0\}$ and $m = +\infty$. Q.E.D.

THEOREM 12. *Let* H, H_i, P_i *and* A *be as in Theorem 10. Assume further that* $(AP_2x, x) \leq -m \|x\|^2$ *for all* $x \in D(A)$ *and some* $m > 0$. *Let* $N: H \longrightarrow H$ *be a potential mapping which has a Gateaux derivative such that*

$$\left(\frac{\partial N}{\partial h}(x), h\right) \geq a\|h\|^2$$

for all $x, h \in H$ *with* $am > 1$ *and* $(\frac{\partial N}{\partial h}(tx), h)$ *is continuous in* t *for* x, h *fixed in* H.

Then the equation $x + ANx = y$ *has at least one solution in* H *for each given* y *in* H.

PROOF. As always it suffices to show that the equation $x+ANx = 0$ has at least one solution in H. Now, as in the proof of Theorem 10, let H' denote the Hilbert space obtained by endowing the linear subspace $D(|A|^{1/2})$ with the inner product

$$[x, y] = (x, y) + (|A|^{\frac{1}{2}}x, |A|^{\frac{1}{2}}y)$$

for $x, y \in D(|A|^{\frac{1}{2}})$. Consider, now, the functional $\Phi: H' \longrightarrow \mathbb{R}$ defined by

$$\Phi(u) = \frac{1}{2}\|P_1u\|^2 - \frac{1}{2}\|P_2u\|^2 + f(|A|^{\frac{1}{2}}u)$$

for $u \in H'$ where $f: H \longrightarrow \mathbb{R}$ is such that grad $f = N$. Now

$$\frac{\partial \Phi}{\partial h}(u) = (P_1 u , h) - (P_2 u , h) + (N|A|^{\frac{1}{2}} u , |A|^{\frac{1}{2}} h)$$

and

$$\frac{\partial^2 \Phi}{\partial k \partial h}(u) = (P_1 k , h) - (P_2 k , h) + (\frac{\partial N}{\partial (|A|^{1/2} k)}(|A|^{\frac{1}{2}} u) , |A|^{\frac{1}{2}} h)$$

for $h, k \in H'$.

So for $u \in H'$ and $h \in H'$ we have

$$\frac{\partial^2 \Phi}{\partial h^2}(u) = \|P_1 h\|^2 - \|P_2 h\|^2 + (\frac{\partial N}{\partial (|A|^{1/2} h)}(|A|^{\frac{1}{2}} u) , |A|^{\frac{1}{2}} h)$$

$$\geqslant \|P_1 h\|^2 - \|P_2 h\|^2 + a \| |A|^{\frac{1}{2}} h\|^2$$

$$\geqslant c|h|^2$$

where

$$c = \min(1, m\varepsilon, \varepsilon) \quad \text{and} \quad a - \frac{1}{m} = 2\varepsilon .$$

Also it follows from our assumptions that $\frac{\partial^2 \Phi}{\partial h^2}(tx)$ is continuous in t for fixed $x \in H'$ and $h \in H'$. It then follows from Theorem 9.4 of Vainberg [27] that there exists a $u_o \in H'$ such that Φ assumes its minimum at u_o. At this point the weak gradient of Φ vanishes and this gives

$$(P_1 u_o - P_2 u_o , h) + (N|A|^{\frac{1}{2}} u_o , |A|^{\frac{1}{2}} h) = 0$$

for all $h \in H' = D(|A|^{\frac{1}{2}})$. This implies that $-N|A|^{\frac{1}{2}}u_o \in D(|A|^{\frac{1}{2}})$ and

$$|A|^{\frac{1}{2}} N|A|^{\frac{1}{2}}u_o = -P_1u_o + P_2u_o.$$

So

$$P_1u_o - P_2u_o + |A|^{\frac{1}{2}} N|A|^{\frac{1}{2}}u_o = 0. \tag{26}$$

It then follows, as in the proof of Theorem 10, that $x_o = |A|^{\frac{1}{2}}u_o$ is a solution of $x + ANx = 0$. This completes the proof of the theorem. Q.E.D.

REMARK 17. We may note that N in Theorem 12 is neither continuous nor bounded demi-continuous in H. However, it is easy to see that N is strongly monotone with constant of strong-monotonicity equal to a. Comparing, now Theorem 12 and 9 we observe that it is the lack of a continuity assumption on N in Theorem 12 which is the cause of non-uniqueness of a solution in Theorem 12 in contrast to Theorem 9. We may, however, add that it is easy to see that the equation

$$P_1u - P_2u + |A|^{\frac{1}{2}} N|A|^{\frac{1}{2}}u = 0$$

in H' (see (26)) has exactly one solution in H', but this does not turn out to be sufficient to obtain the uniqueness of a solution of $x + ANx = 0$. In any case we have the following theorem when A is monotone, without any additional assumptions on N.

THEOREM 13. *Let* H *be a real Hilbert space,* $A: D(A) \longrightarrow H$, *be a linear maximal monotone self-adjoint mapping from domain* $D(A)$ *in* H *to* H. *Let* $N: H \longrightarrow H$ *be a potential mapping which has a Gateaux derivative such that*

$$(27) \qquad (\frac{\partial N}{\partial h}(x)\,,\,h) \geqslant a\|h\|^2$$

for all $x \in H$, $h \in H$ *with* $a > 0$ *and* $(\frac{\partial N}{\partial h}(tx)\,,\,h)$ *is continuous in* t *for* x, h *fixed in* H.

Then the equation $x + ANx = y$ *has exactly one solution in* H *for each given* y *in* H.

PROOF. The existence of a solution of $x + ANx = y$ follows by applying Theorem 12 with $H_1 = H$, $H_2 = \{0\}$ and $m = \infty$.

The uniqueness of a solution of $x + ANx = y$ follows by observing first that the condition (27) implies that

$$(Nx_1-Nx_2\,,\,x_1-x_2) \geqslant a\|x_1-x_2\|^2$$

for $x_1, x_2 \in H$. So if $x_1 + ANx_1 = y = x_2 + ANx_2$ then $x_1-x_2 + ANx_1 - ANx_2 = 0$. Taking inner product with Nx_1-Nx_2 we get

$$0 = (Nx_1-Nx_2\,,\,x_1-x_2) + (Nx_1-Nx_2\,,\,ANx_1-ANx_2) \geqslant a\|x_1-x_2\|^2$$

which imples that $x_1 = x_2$. Hence the solution of $x + ANx = y$ is unique. Hence the theorem. Q.E.D.

THEOREM 14. *Let* H, H_i, P_i *and* A *be as in Theorem 10. Assume further that* $\dim H_2$ *is finite and*

$$(AP_2x , x) \leqslant -m \|x\|^2$$

for all $x \in D(A)$ *and some* $m > 0$. *Let* $N: H \longrightarrow H$ *be the potential of a weakly lower semi-continuous functional* $f: H \longrightarrow \mathbb{R}$ *such that there exists a constant* a *with* $am > 1$ *and*

$$(Nx-Ny , x-y) \geqslant a\|x-y\|^2$$

for all $x, y \in H$. *The the equation* $x + ANx = y$ *has at least one solution in* H *for each given* y *in* H.

PROOF. As always it suffices to show that the equation $x+ANx = 0$ has at least one solution in H. Now, as in the proof of Theorem 10 let H' denote the Hilbert space obtained by endowing the linear subspace $D(|A|^{1/2})$ with the inner product

$$[x , y] = (x , y) + (|A|^{\frac{1}{2}}x , |A|^{\frac{1}{2}}y)$$

for $x, y \in D(|A|^{\frac{1}{2}})$ and let $\Phi: H' \longrightarrow \mathbb{R}$ be the linear functional defined by

$$\Phi(u) = \|P_1u\|^2 - \|P_2u\|^2 + f(|A|^{\frac{1}{2}}u)$$

for $u \in H'$. Then, as in Theorem 10, it follows that Φ is weakly lower semi-continuous on H'. Further, we have for $u \in H'$ that

$$[\operatorname{grad}\Phi(u)\,,\,u] = (P_1u\,,\,u) - (P_2u\,,\,u) + (N|A|^{\frac{1}{2}}u\,,\,|A|^{\frac{1}{2}}u)$$

$$= \|P_1u\|^2 - \|P_2u\|^2 + (N|A|^{\frac{1}{2}}u\,,\,|A|^{\frac{1}{2}}u)$$

$$\geqslant \|P_1u\|^2 - \|P_2u\|^2 + a\|\,|A|^{\frac{1}{2}}u\|^2 - \|NO\|\,\|\,|A|^{\frac{1}{2}}u\|$$

$$\geqslant c|u|^2 - \|NO\|\,|u|$$

where $|u|$ denotes the norm in H',

$$c = \min(1,\, m\varepsilon,\, \varepsilon) \quad \text{and} \quad a - \frac{1}{m} = 2\varepsilon.$$

From this it follows that

$$\Phi(u) - \Phi(0) = \int_0^1 [\operatorname{grad}\Phi(tu)\,,\,u]dt$$

$$= \int_0^1 [\operatorname{grad}\Phi(tu)\,,\,tu]\,\frac{dt}{t}$$

$$\geqslant \frac{c}{2}\,|u|^2 - \|NO\|\,|u| \longrightarrow +\infty$$

as $|u| \longrightarrow +\infty$. So there exists and $R > 0$ such that $\Phi(u) - \Phi(0) > 0$ for $|u| \geqslant R$. Hence the weakly lower semi-continuous functional Φ attains a minimum at a point $u_0 \in H'$ with $|u_0| < R$. This gives that $\Phi(u_0) = 0$ which implies, as in the proof of Theorem 10, that

$$P_1u_o - P_2u_o + |A|^{\frac{1}{2}} N|A|^{\frac{1}{2}}u_o = 0.$$

Applying $(AP_1)^{\frac{1}{2}} - (AP_2)^{\frac{1}{2}}$ to both sides of this last equation and using (24), (25) we then obtain that

$$|A|^{\frac{1}{2}}u_o + AN|A|^{\frac{1}{2}}u_o = 0.$$

This shows that $x = |A|^{\frac{1}{2}}u_o$ is a solution of $x + ANx = 0$ and the proof of the theorem is complete. Q.E.D.

THEOREM 15. *Let* H *be a real Hilbert space,* $A: D(A) \longrightarrow H$ *be a linear maximal monotone self-adjoint mapping from domain* $D(A)$ *in* H *to* H. *Let* $N: H \longrightarrow H$ *be a potential mapping of a weakly lower semi-continuous functional on* H *such that* N *is strongly monotone (i.e., there is a constant* $a > 0$ *such that*

$$(Nx-Ny\,,\, x-y) \geq a\|x-y\|^2$$

for $x, y \in H$).

Then the equation $x + ANx = y$ *has exactly one solution in* H *for each given* y *in* H.

PROOF. Proof of this theorem follows from Theorem 14 in the same way the proof of Theorem 13 followed from Theorem 12.

Northern Illinois University
Department of Mathematics
De Kalb, Illinois
U.S.A.

BIBLIOGRAPHY

[1] BRÉZIS, H., *Équations et inéquations nonlinéiares dans les espaces vectoriels en dualité*, Ann. Inst. Fourier (Grenoble) Vol.18, (1968), pp. 115-175.

[2] BRÉZIS, H., *Inéquations variationelles associées à des opérateurs d'evolution*, Theory and Applications of monotone operators, NATO Summer School, Venice, 1968, Oderisi Gubbio, pp. 1-10.

[3] BRÉZIS, H., *On some degenerate nonlinear parabolic equations*, Proc. Sypos. Pure Math., Vol. 18, Amer. Math. Soc., Providence, R.I. (1970), pp. 28-38.

[4] BRÉZIS, H., *Nonlinear perturbations of monotone operators*, Tech. Report Nº 25, Univ. of Kansas, Lawrence, Kansas (1972).

[5] BROWDER, F.E., *Non linear operators and nonlinear equations of evolution in Banach spaces*, Proc. Sympos. Pure Math., Vol. 18, Part 2, Amer. Math. Soc., Providence, R.I. (1970).

[6] BROWDER, F.E., *Nonlinear functional analysis and nonlinear integral equations of Hammerstein and Urysohn type*, Contributions to Nonlinear Functional Analysis, edited by E.H. Zarantonello, pp. 425-500, Academic Press, New York, 1971.

[7] BROWDER, F.E. and Gupta,C.P., *Monotone operators and nonlinear integral equations of Hammerstein type*, Bull. Amer. Math. Soc., 75 (1969), pp. 1347-1353.

[8] BROWDER, F.E., D.G. de FIGUEIREDO and C.P.GUPTA, *Maximal monotone operators and nonlinear integral equations of Hammerstein type*, Bull. Amer. Math. Soc., 76 (1970), pp. 700-705.

[9] de FIGUEIREDO, D.G. and C.P.GUPTA, *Nonlinear integral equations of Hammerstein type involving unbounded monotone linear mappings*, Jour. Math. Anal. Appl., 39 (1972), pp. 37-48.

[10] de FIGUEIREDO, D.G. and C.P.GUPTA, *Nonlinear integral equations of Hammerstein type with indefinite linear kernel in a Hilbert space*, Indag. Math. **34** (1972), pp.335-344.

[11] de FIGUEIREDO, D.G. and C.P.GUPTA, *On the variational method for the existence of solutions of nonlinear equations of Hammerstein type*, Proc.A.M.S.**40**(1973),pp.470-476.

[12] DOLPH, C.L. and G.J.MINTY, *On nonlinear integral equations of Hammerstein type*, Nonlinear integral equations, edited by P.M. Anselone, Univ. of Wisconsin Press (1964), pp. 99-154.

[13] GUPTA, C.P., *A new existence theorem for nonlinear integral equations of Hammerstein type involving unbounded linear mappings*, (to appear), Proc. Sympos. on Analysis, Rio de Janeiro (1972), Actualités Sci. Ind., Hermann, Paris.

[14] HAMMERSTEIN, A., *Nichtlineare integralgleichungen nebst anwendungen*, Acta Math., 54(1930), pp. 117-176.

[15] KATO, T., *Demi-continuity, hemi-continuity and monotonicity*, Bull. Amer. Math. Soc., 70(1964), pp.548-550.

[16] KATO, T., *Perturbation theory of nonlinear operators*, Springer-Verlag, New York, 1966.

[17] KENMOCHI, N., *Existence theorems for certain nonlinear equations*, Hiroshima Math. Jour., Vol.1 (1971), pp. 435-443.

[18] KOLODNER, I., *Equations of Hammerstein type in Hilbert spaces*, J. Math. Mech., 13 (1964), pp. 701-750.

[19] KOSCIKII, M.E., *Nonlinear equations of Hammerstein type with a monotone operator*, Dokl. Akad. Nauk. USSR, 190 (1970) pp. 31-33 = Soviet Math. Dokl. IV (1970), pp. 25-28.

[20] KRASNOSELSKII, M.A., *Topological methods in the theory of non-linear integral equations*, Pergamon Press, New York, 1964.

[21] LAVRENTIEV, I.M., *On the variational theory of nonlinear equations*, Dokl. Akad. Nauk. USSR, 166 (1966), pp. 284-286 = Soviet Math. Dokl. 8 (1967), pp.993-996.

[22] LAVRENTIEV, I.M., *Solvability of nonlinear equations*, Dokl. Akad. Nauk. USSR, 175 (1967), pp. 1219-1221 = Soviet Math. Dokl., 8 (1967), pp.993-996.

[23] PETRYSHYN, W.V. and P.M.FITZPATRICK, *New existence theorem for nonlinear equations of Hammerstein type*, Transactions Amer. Math. Soc., 160 (1971), pp. 39-63.

[24] PHILLIPS, R.S., *Dissipative operators and hyperbolic systems of partial differential equations*, Transaction Amer. Math. Soc., 90 (1959), pp. 193-254.

[25] ROCKAFELLAR, R.T., *On the maximality of sums of nonlinear monotone operators*, Transactions Amer. Math. Soc., 149 (1970), pp. 75-88.

[26] TRICOMI, F.G., *Integral Equations*, John-Wiley, New York.

[27] VAINBERG, M., *Variational methods for the study of nonlinear operators*, Moscow, 1956. (Engl. transl. Holden-Day, San Francisco, 1964.)

[28] VAINBERG, M., *Nonlinear equations with potential and monotone operators*, Dokl. Akad. Nauk. USSR, 183 (1968), pp. 747-749 = Soviet Math. Dokl. 9 (1968), pp. 1427-1430.

[29] VAINBERG, M. and I.M.LAVRENTIEV, *Equations with monotonic and potential operators in Banach spaces*, Dokl. Akad. Nauk., 187 (1969), pp. 711-714 = Soviet Math. Dokl., 10 (1969), pp. 907-910.

[30] ZAANEN, A.C., *Linear Analysis*, North-Holland Publ. Co., Amsterdam, 1964.

UN'EXTENSION DES THÉORÈMES DE MALGRANGE ET MARTINEAU

par

Domingos Pisanelli

On étudie les operateurs de convolution f de l'espace $H(K_R)$ à valeurs dans $H(K)$ et en suite les équations de convolution $f(y) = 0$.

On obtient comme cas particuliers les théorèmes correspondant de Malgrange et Martineau.

NOTATIONS

$H(\mathbf{C}^n)$ = espace des fonctions holomorphes sur $\mathbf{C}^n$ muni de la famille de seminormes $\|y\|_K = \sup_{t\in K} |y(t)|$ (K compact de $\mathbf{C}^n$).

$H(K)$ = espace des germes de fonctions holomorphes autour d'un compact $K \subset \mathbf{C}^n$, muni de la topologie localement convexe, limite inductive des espaces de Banach $H(\Omega)$ des fonctions holomorphes et bornées sur l'ouvert $\Omega \supset K$.

$|\alpha| = \sqrt{|\alpha_1|^2+\ldots+|\alpha_n|^2}$, $\|\alpha\| = \sup_{1\leqslant j\leqslant n} |\alpha_j|$, $\alpha x = \alpha_1 x_1 + \ldots + \alpha_n x_n$

$\alpha\cdot x = \mathcal{R}(\bar{\alpha}x)$, $\alpha = (\bar{\alpha}_1,\ldots,\bar{\alpha}_n)$, $|\beta| = \beta_1 + \ldots + \beta_n$, $\beta! = \beta_1!\ldots\beta_n!$,

$\alpha^\beta = \alpha_1^{\beta_1}\ldots\alpha_n^{\beta_n}$, $y^{(\beta)} = D_{x_1}^{\beta_1}\ldots D_{x_n}^{\beta_n}$, pour $\alpha = (\alpha_1,\ldots,\alpha_n) \in \mathbf{C}^n$,

$x = (x_1,\ldots,x_n) \in \mathbf{C}^n$ et $\beta = (\beta_1,\ldots,\beta_n) \in \mathbb{N}^n$.

B_R = boule fermée, centrée à l'origine de $\mathbf{C}^n$, pour la norme $|\ |$, de rayon R.

$B_{(R)}$ = boule fermée, centrée à l'origine de $\mathbf{C}^n$, pour la norme $\|\ \|$, de rayon R.

$K_R = K + B_R$

$K_{(R)} = K + B_{(R)}$ (K compact de $\mathbf{C}^n$).

$L(H(\mathbf{C}^n))$ = algèbre des applications linéaires continues de $H(\mathbf{C}^n)$ dans $H(\mathbf{C}^n)$.

exp $\mathbf{C}^n$ = algèbre des fonctions ϕ du type exponentiel (c'est-à-dire ils existent M et A positifs tels que $|\phi(t)| \leqslant M\, e^{A|t|}$ $\forall\, t \in \mathbf{C}^n$).

Exponentielle polynômiale = $p(x)e^{\alpha x}$, où p est un polynôme

L'operateur linéaire et continu f de $H(\mathbf{C}^n)$ dans $H(\mathbf{C}^n)$ est dit de *convolution* si

(1) $f \circ \tau_\alpha = \tau_\alpha \circ f \quad \forall\, \alpha \in \mathbf{C}^n$ où

(a) $(\tau_\alpha y)(x) = y(x + \alpha) \qquad y \in H(\mathbf{C}^n)$.

On dira que "f est permutable avec les operateurs de translation τ_α".

D'une manière équivalente on a

(2) $f \circ D_j = D_j \circ f \qquad (D_j = \frac{\partial}{\partial x_j}\,, \quad 1 \leqslant j \leqslant n)$.

En effet de (a) on obtient

$$\tau_\alpha y = \sum_{|\beta| \geq 0} \frac{y^{(\beta)}}{\beta!} \alpha^\beta \qquad \forall\ \alpha \in \mathbf{C}^n$$

d'où

$$f \circ \tau_\alpha (y) = \sum_{|\beta| \geq 0} \frac{f(y^{(\beta)})}{\beta!} \alpha^\beta$$

$$\tau_\alpha \circ f(y) = \sum_{|\beta| \geq 0} \frac{(f(y))^{(\beta)}}{\beta!} \alpha^\beta .$$

L'identité des séries de puissances nous donne: (1) $\Longleftrightarrow$ (2).

D'une manière équivalente on obtient

(3) $\quad f(e^{\alpha t})(x) = \tilde{f}(\alpha) e^{\alpha x} \qquad \forall\ \alpha,\ x \in \mathbf{C}^n,\ \ \tilde{f}$ étant entière sur $\mathbf{C}^n$.

En effet

$$D_j(f(e^{\alpha t})e^{-\alpha x}) = f(\alpha_j e^{\alpha t})(x) e^{-\alpha x} + f(e^{\alpha t})(x)(-\alpha_j e^{-\alpha x}) = 0$$

pour $1 \leqslant j \leqslant n$, d'où $f(e^{\alpha t})e^{-\alpha x}$ est une fonction analytique de la seule variable α.

Réciproquement nous avons

$$f \circ D_j(e^{\alpha t})(x) = f(\alpha_j e^{\alpha t})(x) = \alpha_j \tilde{f}(\alpha) e^{\alpha x}$$

$$D_j \circ f(e^{\alpha t})(x) = D_j(\tilde{f}(\alpha) e^{\alpha x}) = \tilde{f}(\alpha) \alpha_j e^{\alpha x}.$$

Alors

$$F(e^{\alpha t}) = (D_j \circ f - f \circ D_j)(e^{\alpha t}) = 0.$$

En applicant $(D_\alpha^\beta)_{\alpha=0}$ aux deux membres de la dernière on obtient:

$$F(\alpha^\beta) = 0 \qquad \forall\ \alpha \in \mathbf{C}^n, \quad \beta \in \mathbb{N}^n$$

ce que nous donne que F est nulle sur les polynômes que sont denses dans $H(\mathbf{C}^n)$, d'où $D_j \circ f - f \circ D_j = 0 \quad (1 \leqslant j \leqslant n)$.

Observons que $\tilde{f}$ est de type exponentiel. En effet la continuité et la linéarité de f nous donnent, étant donné K compact de $\mathbf{C}^n$, qu'ils existent H compact de $\mathbf{C}^n$ et $M > 0$ convenables tels que

$$\|f(y)\|_K \leqslant M \|y\|_H ,$$

d'où

$$|f(e^{\alpha t})(x)| \leqslant M\, e^{\sup\limits_{t \in H} \mathcal{R}(\alpha t)} \qquad \forall\ x \in K,\ \alpha \in \mathbf{C}^n,$$

$$|\tilde{f}(\alpha)| \leqslant M\, e^{\sup\limits_{t \in H} \mathcal{R}(\alpha t) - \mathcal{R}(\alpha x)} = M\, e^{\bar{\alpha}\cdot t_1 - \bar{\alpha}\cdot x}$$

$$|\tilde{f}(\alpha)| \leqslant M\, e^{|\alpha|\,|t_1 - x|}$$

étant t_1 convenable de H. On obtient

$$|\tilde{f}(\alpha)| \leqslant M\, e^{|\alpha| A} \qquad \forall\ \alpha \in \mathbf{C}^n,$$

$(A = \sup\limits_{H \times K} |t-x|)$.

Supposons maintenant que le type de $\tilde{f}$ soit $\leqslant R$. On aura pour: $\varepsilon > 0 \longrightarrow M_\varepsilon > 0$ convenable

$$|\tilde{f}(\alpha)| \leqslant M_\varepsilon e^{|\alpha|(R+\varepsilon)} \qquad \forall\ \varepsilon > 0,\ \alpha \in \mathbf{C}^n$$

ou

$$|\tilde{f}(\alpha)| \leqslant M_\varepsilon e^{\sup_{B_R} \mathcal{R}(\alpha t)+\varepsilon|\alpha|} .$$

Du théorème de Martineau ([2] pg. 151, th. 6) on obtiendra qu'il existe $T_\gamma \in H'(B_R)$ tel que $\hat{T}_\gamma(\alpha) = T_\gamma(e^{\alpha\gamma}) = \tilde{f}(\alpha)$. On aura

(4) $$f(y)(x) = T_\gamma\ y(x+\gamma) \qquad y \in H(\mathbf{C}^n), \quad T_\gamma \in H'(B_R).$$

En effet

$$f(e^{\alpha t})(x) = \tilde{f}(\alpha)e^{\alpha x}$$

et

$$T_\gamma(e^{\alpha(x+\gamma)}) = e^{\alpha x}\ T_\gamma(e^{\alpha\gamma}).$$

En raisonnant comme dans la démonstration de (3) nous obtenons (4).

(4) $\Longrightarrow$ (3) est immédiate.

(5) L'application $f \in L(H(\mathbf{C}^n)) \longmapsto \tilde{f} \in \exp \mathbf{C}^n$ est un isomorphisme d'algèbre. Cett'application est "sur" par la propriété (4).

Elle est biunivoque parce que elle est determinée par les valeurs de f sur les exponentielles. Un calcul immédiat nous donne

$$\widetilde{f_1 \circ f_2} = \tilde{f}_1 \circ \tilde{f}_2 .$$

Si $y \in H(K_{(R)}) \hookrightarrow H(K_R)$ on aura

(6) $$f(y) = \sum_{|\beta| \geqslant 0} \frac{y^{(\beta)}}{\beta!} a_\beta \qquad a_\beta \in \mathbf{C}.$$

En effet

$$y(x+\gamma) = \sum_{|\beta| \geqslant 0} \frac{y^{(\beta)}(x)}{\beta!} \gamma^\beta ,$$

uniforment quand x est autour de K e γ autour de $B_{(R)}$ (inégalité de Cauchy). La formule (6) est alors une consequence immédiate de (4).

La formule (4) nous défine un'application lineaire et continue f_* sur un $H(K_R)$ quelconque à valeurs dans $H(K)$. Soient I et J les applications canoniques de $H(\mathbf{C}^n)$ dans $H(K_R)$ et de $H(\mathbf{C}^n)$ dans $H(K)$.

On obtient le diagramme commutatif

$$\begin{array}{ccc} H(\mathbf{C}^n) & \xrightarrow{\quad f \quad} & H(\mathbf{C}^n) \\ {\scriptstyle I}\downarrow & & \downarrow{\scriptstyle J} \\ H(K_R) & \xrightarrow[\quad f_* \quad]{} & H(K) \end{array}$$

Quand K_R est de Runge (c'est-à-dire les polynomes sont denses dans $H(K_R)$), f_* est unique.

On prendra dorénavant K_R *de Runge* et comme operateur de convolution

$$(7) \qquad f(y)(x) = T_\gamma \, y(x+\gamma) \qquad y \in H(K_R), \quad T_\gamma \in H(B_R).$$

II) LEMME D'AVANISSIAN ([2], pg. 133):

Soit G *une fonction entière de la variable complexe* z *de type expoentiel* $\leqslant R$:

$$|G(z)| \leqslant M_\varepsilon \, e^{(R+\varepsilon)|z|} \qquad \forall \varepsilon > 0, \qquad z \in \mathbf{C}.$$

Alors

$$M\,(\log |G|, z, \lambda|z|) \geqslant \left(\frac{1+\lambda}{\lambda}\right)^2 \log |G(0)| +$$

$$+ \left[1 - \left(\frac{1+\lambda}{\lambda}\right)^2\right](1+\lambda)(R+\varepsilon)|z|$$

pour tout $|z| > \frac{2}{\varepsilon} \cdot \log M_\varepsilon$ *et* $\lambda > 0$.

Nous indicons ici par $M(\phi, z, R)$ la moyenne de la fonction ϕ, d'une variable complexe, sur le disque de centre z et rayon R:

$$\frac{1}{\pi R^2} \iint_{|w| \leqslant R} \phi(z+w)\, dx\, dy \qquad z \in \mathbf{C}, \quad w = x + iy.$$

THÉORÈME 1. *Soient* K_1 *convexe compact de* $\mathbf{C}^n$, $X \in H'(K_1)$, $f \neq 0$ *de convolution et* $\hat{X}/\tilde{f}$ *entière. Il existe alors* $T_1 \in H'(K_1+B_d)$ *tel que* $\hat{X}/\tilde{f} = \hat{T}_1$ *où*

$$d = \inf_{\lambda>0} \{ \sup_{\substack{|u|\leq 1 \\ t\in K_1}} R(ut) + (\frac{1}{\lambda}+1)(\frac{1}{\lambda}+2)R\}.$$

DÉMONSTRATION. La fonction $\log|\hat{X}/\tilde{f}(zv)|$ est sous-harmonique de la variable complexe z (ayant fixé v, $|v| = 1$):

$$(8) \qquad \log |\hat{X}/\tilde{f}(zv)| \leq M(\log |\hat{X}(\zeta v)|, z, \lambda|z|) - M(\log |\tilde{f}(\zeta v)|, z,\lambda|z|)$$

La caracterisation de Martineau ([2], pg. 151), des formes linéaires et continues sur $H(K_1)$ (K_1 convexe compact) nous donne que

$$|\hat{X}(u)| \leq M_\varepsilon\, e^{\sup_{t\in K_1} R(ut)+\varepsilon|u|} \qquad \forall\, \varepsilon > 0, \quad u\in \mathbf{C}^n$$

d'où

$$\log |\hat{X}(u)| \leq \log M_\varepsilon + \sup_{t\in K_1} R(ut) + \varepsilon|u|$$

et

$$M(\log |\hat{X}(\zeta v)|, z,\lambda|z|) \leq \log M_\varepsilon + \sup_{t\in K_1} R(zvt) +$$

$$+ \lambda|z| \sup_{t\in K_1} R(vt) + \varepsilon|z|(1 + \lambda).$$

Le lemme d'Avanissian et (8) nous donnent:

$$\log |\hat{X}/\tilde{f}(u)| \leq \log M_\varepsilon + \sup_{t\in K_1} R(ut) +$$

$$+ |u|(\lambda \sup_{\substack{t\in K_1 \\ |u|\leq 1}} R(ut) + (1+\frac{1}{\lambda})(2+\frac{1}{\lambda})R) +$$

$$+ \varepsilon|u|(1+\lambda) + \varepsilon|u|(1+\frac{1}{\lambda})(2+\frac{1}{\lambda}) - (\frac{1+\lambda}{\lambda})^2 \log|\tilde{f}(0)|.$$

Ayant donné $\varepsilon' > 0$, on prendra $\lambda > 0$ tel que

$$\lambda \sup_{\substack{t\in K_1 \\ |u|\leq 1}} R(ut) + (1+\frac{1}{\lambda})(2+\frac{1}{\lambda})R < d + \frac{\varepsilon'}{3} ,$$

$$\varepsilon < \frac{\varepsilon'}{3} \cdot \frac{1}{1+\lambda}$$

$$\varepsilon < \frac{\varepsilon'}{3} \cdot \frac{1}{(1+\frac{1}{\lambda})(2+\frac{1}{\lambda})} .$$

À la fin on obtient:

$$\log|\hat{X}/\tilde{f}(u)| \leq A(\varepsilon') + \sup_{t\in K_1} R(ut) + |u|(d+\varepsilon') =$$

$$= A(\varepsilon') + \sup_{K_1+B_d} R(ut) + |u|\varepsilon'$$

et par le théorème de Martineau ([2], pg. 151, th. 6)

$$\hat{X}/\tilde{f} = \hat{T}_1 \qquad T_1 \in H'(K_1+B_d).$$

Quand $\tilde{f}(0) = 0$, en supposant $\tilde{f} \neq 0$, il existira $u_o \in \mathbf{C}^n$ tel que $\tilde{f}(u_o) \neq 0$. On obtiendra

$$|\hat{X}(u+u_o)/\tilde{f}(u+u_o)| \leq M_{\varepsilon'} \, e^{\sup_{K_1+B_d} R(ut)+\varepsilon'|u|} \qquad \forall\ \varepsilon' > 0,\ u \in \mathbf{C}^n$$

d'où

$$|\hat{X}(u)/\tilde{f}(u)| \leqslant N_{\varepsilon'} \, e^{\sup_{K_1+B_d} R(ut)+\varepsilon'|u|} \qquad \forall\, \varepsilon' > 0, \quad u \in \mathbf{C}^n.$$

THÉORÈME 2. *Soit* $f \neq 0$ *de convolution entre* $H(K_R)$ *et* $H(K)$ *et supposons* K_1 *convexe compact tel que* $K_1+B_d \subset K$. *Les solutions de l'équation* $f(y)=0$ *sont dans l'adhérence dans* $H(K_1)$, *du sous-espace engendré par les solutions exponentielles polynômiales.*

DÉMONSTRATION. Soit $X \in H'(K_1)$ nulle sur les solutions exponentielles polynômiales de l'équation $f(y) = 0$.

Démonstrons que $\hat{X}/\tilde{f}$ est entière. En effet on a

$$f(e^{\alpha t})(x) = \tilde{f}(\alpha)e^{\alpha x}.$$

Si on suppose $\tilde{f}(\alpha_o) = 0$ on aura $f(e^{\alpha_o t}) = 0$ et

$$\hat{X}(\alpha_o) = X(e^{\alpha_o t}) = 0.$$

En comptant la multiplicité de la racine α_o de $\tilde{f}(\alpha) = 0$, sur les droites par α_o, on obtient (Malgrange-Hörmander) que α_o est une racine de $\hat{X}(\alpha) = 0$ de multiplicité non inferieure de celle de α_o dans l'équation $\tilde{f}(\alpha) = 0$.

Il existirá, par le théorème 1, $T_1 \in H'(K_1+B_d)$ tel que $\hat{X}/\tilde{f} = \hat{T}_1$ d'où

$$\widehat{XoI} = \hat{X} = \widehat{T_1 oJof},$$

I et J étant les applications canoniques de $H(K)$ dans $H(K_1+B_d)$ et de $H(K_R)$ dans $H(K_1)$:

$$H(K_R) \xrightarrow{f} H(K) \xrightarrow{I} H(K_1+B_d) \xrightarrow{T_1} \mathbf{C}$$

$$H(K_R) \xrightarrow{J} H(K_1) \xrightarrow{X} \mathbf{C} .$$

K_R est de Runge par hypothèse, on aura

$$X \circ J = T_1 \circ I \circ f.$$

Si y est racine de $f(y) = 0$, on aura $X(J(y)) = 0$. Le théorème de Hahn-Banach nous donne que les solutions de $f(y) = 0$ sont, dans l'adhérence en $H(K_1)$ du sous-espace des solutions exponentielles polynomiales.

CAS PARTICULIERS

1) Quand $R = 0$ on obtient $d = 0$. Si on prend $K_1 = K$ on aura le théorème de Martineau: *Les solutions de l'équation de convolution* $f(y) = 0$, $f: H(K) \longrightarrow H(K)$, *sont dans l'adhérence du sous-espace engendré par les exponentielles polynomiales* ([3], pg. 314).
2) Quand y est solution entière de l'équation de convolution $f(y)=0$, $f: H(\mathbf{C}^n) \longrightarrow H(\mathbf{C}^n)$, on aura $y \in H(K_R)$ $\forall$ K. Alors y *peut être approchée, sur tout compact convexe* $K_1 \subset \mathbf{C}^n$, *par des combinaisons linéaires d'exponentielles polynomiales* ([1]).
3) Quand $K_1 = \{0\}$ on aura $d = \inf_{x>0} (x+1)(x+2)R = 2R$. *Chaque solution de l'équation* $f = 0$, $f: H(B_{3R}) \longrightarrow H(B_{2R})$, *est dans l'adhérence en* $H(\{0\})$ *du sous-espace engendré par les solutions exponentielles polynomiales.*

ADDENDA

On peut étudier directement les operateurs de convolution

$$f: H(K_R) \longrightarrow H(K),$$

K étant compact connexe et K_R de Runge compact.

On dira que f *est de convolution* si f est linéaire, continu et

$$D_j \circ f = f \circ D_j \qquad 1 \leq j \leq n,$$

$D_j = \frac{\partial}{\partial z_j}$ étant l'operateur de derivation de $H(K_R)$ ou de $H(K)$.

D'une manière équivalente on obtient la propriété

$$\tau_\alpha \circ f = f \circ \tau_\alpha .$$

τ_α étant l'operateur des "petites" translation ([3], pg. 313).

On obtient aussi

$$f(e^{\alpha t})(x) = \tilde{f}(\alpha) e^{\alpha x}.$$

Maintenant on peut montrer que $\tilde{f}(\alpha)$ est de type exponentiel $\leq R$. En effet l'image par f de la boule unitaire de $H(K_R+B_\varepsilon)$ $(\varepsilon > 0)$ est bornée. Ils existent alors un ouvert $H(\varepsilon) \supset K$ et $M > 0$ tels que

$$\| f(y) \|_{H(\varepsilon)} \leq M$$

dès que $\|y\|_{K_R+B_\varepsilon} \leq 1$. Alors

$$|f(e^{\alpha t})(x)| \leq M \, e^{\sup_{K_R+B_\varepsilon} \mathcal{R}(\alpha t)} \qquad x \in K$$

d'où

$$|\tilde{f}(\alpha)| \leq M \, e^{\sup_K \mathcal{R}(\alpha t) + R|\alpha| + \varepsilon|\alpha| - \mathcal{R}(\alpha x)}$$

et

$$|\tilde{f}(\alpha)| \leq M\, e^{\sup_K \mathcal{R}(\alpha t) + (R+\varepsilon)|\alpha| - \sup_K \mathcal{R}(\alpha x)} .$$

On obtient aussi:

$$f(y)(x) = T_\gamma(y(x+\gamma)) \qquad y \in H(K_R), \quad T_\gamma \in H'(B_R)$$

que nous donne: *Si* y *est une fonction entière alors* f(y) *se prolonge en une fonction entière.*

On a de même:

$$f(y) = \sum_{|\beta| \geq 0} \frac{a_\beta}{\beta!} y^{(\beta)} \qquad y \in H(K_{(R)}), \quad a_\beta \in \mathbf{C}.$$

Universidade de S.Paulo
Instituto de Matemática e Estatística
Caixa Postal 20570
S.Paulo
BRASIL

BIBLIOGRAPHIE

[1] MALGRANGE, B., *Existence et approximation des solutions des équations aux derivées partielles et des équations de convolution*, Ann. Inst. Fourier, Grenoble, t.6 (1955-56), pg. 271-355.

[2] MARTINEAU, A., *Équations differentielles d'ordre infini.* Boll. Soc. Math. France, 95 (1967), pg. 109-154.

[3] PISANELLI, D., *Sullo sviluppo di Pincherle di un operatore lineare.* An. Acad. Bras. Ciências (1971), 43 (2).

[4] NACHBIN, L., "Quinzena de Análise Funcional", São José dos Campos, 1969.

PARTIAL DIFFERENTIAL EQUATIONS IN HOLOMORPHIC FOCK SPACES

by

Thomas A. W. Dwyer, III

INTRODUCTION. We wish to describe here some existence theorems on partial differential equations in spaces of holomorphic mappings with function space domains. When the domains are the spaces $L^2(\mathbb{R})$ of square-integrable functions we obtain the holomorphic Fock space of [Be]. When the domains are the spaces $S(R)$ and $S'(R)$ of rapidly decreasing functions and tempered distributions we obtain spaces of entire functions which are representations of tempered distributions in inficinte dimension in the sense of [KMP]. Details of proofs, in the abstract setting of holomorphic mappings on projective and inductive limits of Hilbert spaces, will appear elsewhere [D2, D3].

HOLOMORPHIC FOCK SPACES OVER $L^2(\mathbb{R})$. Holomorphic Fock space over $L^2(\mathbb{R})$ is the Hilbert space $F(L^2(\mathbb{R}))$ of entire functions

$$F: L^2(\mathbb{R}) \longrightarrow C$$

of the form

$$(+) \qquad F(f) = \sum_{n=0}^{\infty} \frac{1}{n!} F_n(f)$$

with polynomial terms

$$F_n(f) = \int \cdots \int f_n(x_1, \ldots, x_n) f(x_1) \ldots f(x_n) dx_1 \ldots dx_n$$

given by kernels f_n in $L_s^2(\mathbb{R}^n)$ (symmetric functions), with respect to the scalar product

$$(*) \quad ((F \mid F)) = |||F|||^2 = \sum_{n=0}^{\infty} \frac{1}{n!} \int \cdots \int f_n(x_1, \ldots, x_n) \overline{f_n(x_1, \ldots, x_n)} dx_1 \ldots dx_n.$$

REMARK. $F(L^2(\mathbb{R}))$ is a state space for quantum systems with an infinite number of degrees of freedom, and differential operators on it describe observables of the systems. Differential operators on the analogous space $F(\mathbb{C}^n)$ with finite-dimensional domain have been studied in [T] and [NS], and group representations in [Ba].

The F_n in (+) are called homogeneous Hilbert-Schmidt polynomials, and the F entire functions of Hilbert-Schmidt type, on $L^2(\mathbb{R})$. In general, when the derivative polynomials of a holomorphic function on $L^2(\mathbb{R})$ at a point f have function kernels, these kernels are denoted by

$$(x_1, \ldots, x_n) \longmapsto \frac{\delta^n F(f)}{\delta f(x_1) \ldots \delta f(x_n)}$$

called Fréchet-Volterra variational derivatives of F at f [DL]. In G. Coeuré's notation we have

$$d_f^n F(h) = \int \cdots \int \frac{\delta^n F(f)}{\delta f(x_1) \ldots \delta f(x_n)} h(x_1) \ldots h(x_n) dx_1 \ldots dx_n ,$$

with h in $L^2(\mathbb{R})$. It turns out that $(*)$ finite implies the derivatives of any F in $F(L^2(\mathbb{R}))$ at *all* f in $L^2(\mathbb{R})$ have square-integrable kernels [Dl].

PARTIAL DIFFERENTIAL OPERATORS. Given a Hilbert-Schmidt polynomial

$$P = \sum_{n=0}^{m} P_n$$

on $L^2(\mathbb{R})$, with $P_n(f) = (f^{\otimes n} \mid p_n)_{L^2}$, p_n in $L_s^2(\mathbb{R}^n)$, the differential operator $P(d)$ is the sum of the operators $P_n(d)$ of the form

$$P_n(d)F(f) = \int \ldots \int \frac{\delta^n F(f)}{\delta f(x_1) \ldots \delta f(x_n)} p_n(x_1, \ldots, x_n) dx_1 \ldots dx_n .$$

If $p_n(x_1, \ldots, x_n)$ is the symmetric product of $f_1(x_1), \ldots, f_n(x_n)$ with f_k in $L^2(\mathbb{R})$ then

$$P_n(d) = \frac{\partial^n}{\partial f_1 \ldots \partial f_n}$$

(directional derivatives). As in [D1], we have:

PROPOSITION. $((P(d)F \mid G)) = ((F \mid \overline{P} \cdot G))$.

An extension to Hilbert-Schmidt polynomials of an estimate of Trèves on polynomials in $\mathbf{C}^n$ [T, D1], yields the following universal estimate:

$$(**) \qquad |||P \cdot F||| \geq m!^{\frac{1}{2}} \, \|p_m\|_{L^2} \, |||F||| .$$

The consequent continuity of division by polynomials and the adjointness of $P(d)$ and P yields the following existence theorem, given in [D1]:

THEOREM. $P(d)F(L^2(\mathbb{R})) \supset F(L^2(\mathbb{R}))$.

However, $P(d)F(L^2(\mathbb{R})) \neq F(L^2(\mathbb{R}))$. One approach to have $P(d)$ continuously defined everywhere on an appropriate domain is to introduce weight sequences $(s_n)_n = s$ and appropriate dual sequences $(s_n')_n = s'$ (Stufenräume), defining spaces $F_s(L^2(\mathbb{R}))$ with coefficients $s_n \cdot (n!)^{-1}$ in place of $\frac{1}{n!}$ in (+), and taking limits $\text{inv lim}_s\ F_s(L^2(\mathbb{R}))$, $\text{dir lim}_{s'}\ F_{s'}(L^2(\mathbb{R}))$ by properly ordering the sequences. Inclusion and existence theorems for $s_n = r^n$, $r > 0$ are obtained by modifying the "a priori" estimate (**) to the norms of weighted spaces [Dl]. These spaces have $\mathbf{C}^n$-analogues in [M]. Weights $s_n = (n+1)^r$ in turn extend to Hilbert space domains the spaces of entire functions giving holomorphic representations of distributions studied in [Ba]. Since $L^2(\mathbb{R})'$ is contained in each $F_s(L^2(\mathbb{R}))$ topologically, none of these spaces are nuclear [Dl].

HOLOMORPHIC FOCK SPACES OVER $S(\mathbb{R})$ AND $S'(\mathbb{R})$. Another approach, yielding $P(d)$ continuous on nuclear spaces of holomorphic mappings, is to restrict the domain from $L^2(\mathbb{R})$ to the rapidly decreasing functions $S(\mathbb{R})$ or extend it to the tempered distributions $S'(\mathbb{R})$, obtaining "Fock spaces" $F(S(\mathbb{R}))$ of entire functions $F: S(\mathbb{R}) \longrightarrow \mathbf{C}$ as in (+) but with f_n in $S_s'(\mathbb{R}^n)$ (symmetric distributions), as well as "Fock spaces" $F(S'(\mathbb{R}))$ of maps $F: S'(\mathbb{R}) \longrightarrow \mathbf{C}$ with f_n in $S_s(\mathbb{R}^n)$ (symmetric functions). For this one decomposes $S(\mathbb{R})$ as a projective limit of Sobolev-like Hilbert spaces $S_r(\mathbb{R})$ of functions f in $L^2(\mathbb{R})$ such that $(f \mid h^r f)_{L^2} < \infty$, $r \geqslant 0$, where $h = (x + d/dx)(x - d/dx)$ is essentially the Hermite operator [BS, KMP], defining

$$F(S'(\mathbb{R})) = \text{inv lim}_r\ F(S_r'(\mathbb{R}))$$

and

$$F(S(\mathbb{R})) = \text{dir lim}_r\, F(S_r(\mathbb{R})).$$

Here $F(S_r(\mathbb{R}))$ and $F(S_r'(\mathbb{R})) = F(S_{-r}(\mathbb{R}))$ are defined as $F(L^2(\mathbb{R}))$, with the scalar product (*). The n-homogeneous polynomials on $S(\mathbb{R})$ are given by symmetric tempered distribution kernels in $\mathbb{R}^n$, and on $S'(\mathbb{R})$ by symmetric rapidly decreasing function kernels. All elements in $F(S(\mathbb{R}))$ are clearly entire functions. That all elements of $F(S'(\mathbb{R}))$ are entire is a consequence of the maps $S_r'(\mathbb{R}) \longrightarrow S_p'(\mathbb{R})$ being compact for r sufficiently larger than p, making $S'(\mathbb{R})$ a "Silva space" (dual of a Fréchet-Schwartz space) [P].

PROPOSITION. $F(S'(\mathbb{R}))$ *and* $F(S(\mathbb{R}))$ *are in duality for the bilinear form* $\ll\, ,\, \gg$ *defined by the inner product* (*). *This is proved, for example, by means of the Fourier-Borel transformation* cf. [Dl, Mt].

Polynomials on $S(\mathbb{R})$ give rise to operators $P'(d)$ on $F(S'(\mathbb{R}))$, and polynomials P on $S'(\mathbb{R})$ to operators $P(d)$ on $F(S(\mathbb{R}))$, defined as on $F(L^2(\mathbb{R}))$ but now in the sense of generalized functions.

Estimates of the type

$$|||P_m(d)F'|||_r \leqslant \text{const.}\, \|P_m\|_{S_r}\, |||F'|||_{r-1}\,,$$

for F' in $F(S(\mathbb{R}))$ and the norm $|||\ |||_r$ on $F(S_r(\mathbb{R}))$, and similarly on $F(S'(\mathbb{R}))$, lead to the desired continuous inclusions:

PROPOSITION. $P(d)F(S(\mathbb{R})) \subset F(S(\mathbb{R}))$ *for any continuous polynomial* P *on* $S'(\mathbb{R})$, *and* $P'(d)F(S'(\mathbb{R})) \subset F(S'(\mathbb{R}))$ *for any continuous polynomial* P' *on* $S(\mathbb{R})$.

We are also able to obtain the $P(d) \longleftrightarrow P$ duality:

PROPOSITION. $\langle\langle P'(d), G'\rangle\rangle = \langle\langle F, P'\cdot G'\rangle\rangle$ *and* $\langle\langle P\cdot F, G'\rangle\rangle = \langle\langle F, P(d)G'\rangle\rangle$ *for* F *in* $F(S'(\mathbb{R}))$, G' *in* $F(S(\mathbb{R}))$, P *a continuous polynomial on* $S'(\mathbb{R})$ *and* P' *on* $S(\mathbb{R})$.

Projective and inductive limit techniques, a lemma on extensions of factors of products of Hilbert-Schmidt polynomials from $S(\mathbb{R})$ to $S_r(\mathbb{R})$, and the estimate (**) on $S_r(\mathbb{R})$ and $S_{-r}(\mathbb{R})$ instead of $L^2(\mathbb{R})$ lead to continuity of division by polynomials, and through transposition and the Hahn-Banach theorem to new existence results:

THEOREM. $P'(d)F(S'(\mathbb{R})) = F(S'(\mathbb{R}))$ and $P(d)F(S(\mathbb{R})) = F(S(\mathbb{R}))$ for *every continuous polynomial* P' *on* $S(\mathbb{R})$ *and* P *on* $S'(\mathbb{R})$. *For details we refer to* [D3].

REMARK. Identifying $F(S'(\mathbb{R}))$ and $F(S(\mathbb{R}))$ with suitable completions of $\mathbf{C}\oplus S(\mathbb{R})\oplus S_s(\mathbb{R}^2)\oplus\ldots$ and $\mathbf{C}\oplus S'(\mathbb{R})\oplus S_s'(\mathbb{R}^2)\oplus\ldots$ leads to showing that $F(S'(\mathbb{R}))$ and $F(S'(\mathbb{R}))$ are respectively isomorphic to an abstract space $\mathfrak{G}^*$ of "tempered distributions" and the corresponding space of "test functions" in infinite dimension of [KMP]. The "annihilation" operators are realized by directional derivatives and the "creation" operators by multiplication by linear functionals. Still another isomorphic realization of such spaces, for which the results above also hold, is with $L^2(\mathbb{R})$ replaced by the "Fischer space" $F(\mathbf{C})$ and $S_r(\mathbb{R})$ by the space

$$F_r(\mathbf{C}) = \{f \in H(\mathbf{C}) \mid \frac{1}{\pi}\iint |f(z)|^2(1+|z|^2)^r \exp(-|z|^2)\,dxdy < \infty\}$$

(where $z = x + iy$) of [Ba].

Northern Illinois University
Department of Mathematics
De Kalb, Illinois
U.S.A.

REFERENCES

[Ba] V.BARGMANN, *On a Hilbert Space of Analytic Functions and an Associated Integral Transform, Part I*, Commun. Pure Appl. Math. 14(1961), 187-214, *Part II*, Ibid. 20 (1967), 1-101.

[Be] BEREZIN, *The Method of Second Quantization*, Academic Press, New York, 1966.

[BS] B.SIMON, *Distributions and Hermite Expansions*, J.Math. Phys. 12(1971), 140-148.

[D1] T.DWYER, *Partial Differential Equations in Fischer-Fock Spaces for the Hilbert-Schmidt Holomorphy Type*, Bull. Amer. Math. Soc. 77(1971), 725-730.

[D2] T.DWYER, *Holomorphic Representations of Tempered Distributions and Weighted Fock Spaces*, Proceedings of the Colloquium on Analysis, Universidade Federal do Rio de Janeiro, 15-24 August 1972, to appear.

[D3] T.DWYER, *Holomorphic Fock Representations and Partial Differential Equations on Countably Hilbert Spaces*, Bull.Amer.Math.Soc. 79(1973),1045-1050.

[DL] M.DONSKER and J.LIONS, *Fréchet-Volterra Variational Equations, Boundary Value Problems and Function Space Integrals*, Acta Math. 108(1962), 147-228.

[KMP] P.KRISTENSEN, L.MEJLBO and E.POULSEN, *Tempered Distributions in Infinitely Many Dimensions, I: Canonical Field Operators*, Commun. Math. Phys. 1(1965), 175-214.

[M] A.MARTINEAU, *Équations Différentielles d'Ordre Infini*, Bull. Soc. Math. France 95(1967), 109-154.

[Mt] M.MATOS, *Thesis*, University of Rochester, Rochester, New York, 1971.

[NS] D.NEWMAN and H.SHAPIRO, *Certain Hilbert Spaces of Entire Functions*, Bull. Amer. Math. Soc. 72(1966), 917-977.

[P] D.PISANELLI, *Sur les Applications Analytiques en Dimension Infinie*, C.R.Acad. Sci. Paris 274(1972), 760-762.

[T] F.TRÈVES, *Linear Partial Differential Equations with Constant Coefficients*, Gordon and Breach, New York, 1966.

A NOTE ON THE EIGENVALUES OF COMPACT OPERATORS

by

R. Ramalho

INTRODUCTION

In this paper we shall study the existence of eigenvalues and eigenfunctions of the equation $\lambda x = Tx$, where T is a compact (not necessarily linear) operator on a subset of a Banach space X.

In Section 1 we shall present an abstract result (Theorem 1) while in Section 2 we give two applications of this result to the eigenvalue problem for systems of Hammerstein integral equations. One of these applications contains the following theorem of S.Karlin and L.Nirenberg (see [2]):

THEOREM. *Let* ϕ *be a continuous linear functional defined on* $C_o[0,1]$. *Let* $K(s,t)$ *be a continuous function defined on the closed unit square* $0 \leqslant s,t \leqslant 1$ *such that for every* t:

$$\phi(K(\cdot,t)) > 0, \quad 0 \leqslant t \leqslant 1$$

and let

$$a = \min_{s,t} \frac{K(s,t)}{\phi(K(\cdot,t))}, \quad b = \max_{s,t} \frac{K(s,t)}{\phi(K(\cdot,t))} .$$

Suppose $f(t,\xi)$ *is a continuous positive function defined on* $0 \leqslant t \leqslant 1$, $a \leqslant \xi \leqslant b$. *Then there exists a continuous function* $v(t)$ *with* $\phi(v) = 1$ *satisfying:*

$$\int_0^1 K(s,t)f(t,v(t))dt = \lambda v(s), \quad 0 \leqslant s \leqslant 1$$

for some constant $\lambda > 0$.

Finally in Section 3 we prove Theorem 2, which extends in some sense the abstract result of Section 1.

We wish to express our gratitude to L. Nirenberg and H. Brézis for their patience in listening to our ideas on the subject and for their helpful suggestions.

SECTION 1

Here we shall prove the following

THEOREM 1. *Let* X *be a Banach space and* $\phi: X \longrightarrow \mathbb{R}$ *be a continuous linear functional. Consider the affine closed hyperplane:*

$$H = \{x \in X \mid \phi x = 1\}$$

and suppose that $T: H \longrightarrow X$ *is a continuous compact operator.*

If, for every $x \in H$, *the following conditions are satisfied:*

i) $\phi(Tx) > 0$

ii) $\exists\, M > 0$ *such that* $\|Tx\| \leqslant M[\phi(Tx)]$,

then there exists $\lambda_o > 0$ *and there exists* $x_o \in H$ *with* $\lambda_o x_o = Tx_o$.

PROOF. Let $\overline{B} = \{x \in X \mid \|x\| \leqslant M\}$ and assume that the closed bounded convex set $A = H \cap \overline{B}$ is non-empty. Define the operator:

$$S: A \longrightarrow X$$

$$x \longmapsto Sx = \frac{Tx}{\phi(Tx)}$$

which is clearly continuous and compact on A.

Moreover $S(A) \subset A$. In fact let $y = Sx$ be any element of A. Then

$$\phi y = \phi(Sx) = \phi \frac{Tx}{\phi(Tx)} = 1,$$

so that $y \in H$, and, on the other hand,

$$\|y\| = \|Sx\| = \frac{\|Tx\|}{\phi(Tx)} \leqslant M$$

so that $y \in \overline{B}$.

Now a direct application of the Schauder fixed point theorem yields the existence of $x_o \in A$ with

$$x_o = Sx_o = \frac{Tx_o}{\phi(Tx_o)},$$

which proves Theorem 1, with $\lambda_o = \phi(Tx_o) > 0$. We also conclude that $\|x_o\| \leqslant M$.

If $A = H \cap \overline{B}$ is empty, we can always find $M_1 \geqslant M$ such that the closed bounded convex set $A_1 = H \cap \overline{B}_1$ is non-empty, where

$$\overline{B}_1 = \{x \in X \mid \|x\| \leqslant M\}.$$

In fact, if there exists at least one $x_1 \in \overline{B}$ such that $\phi x_1 = \alpha$, with $|\alpha| > 1$, then $y_1 = \frac{1}{\alpha} x_1$ clearly belongs to A_1. If, however, $|\phi x| < 1$ for all $x \in \overline{B}$, then let $x_2 \in \overline{B}$ be such that $\phi x_2 = \beta$ with $|\beta| < 1$ and take $y_2 = \frac{1}{\beta} x_2$. It follows that $\phi y_2 = 1$ and that

$$\|y_2\| = \frac{1}{|\beta|} \|x_2\| \leqslant \frac{1}{|\beta|} M,$$

hence $y_2 \in A_1$, if $M_1 = \frac{M}{|\beta|}$.

SECTION 2

In this section we shall work in the Banach space $C_0^n[0,1]$ of continuous vector-valued functions $u(s) = (u_1(s), u_2(s), \dots, u_n(s))$ with the norm

$$\|u\| = \sum_{i=1}^{n} \max_{s} |u_i(s)|$$

and consider the eigenvalue problem for a system of Hammerstein integral equations of the form

$$\lambda u_i(s) = \sum_{j=1}^{n} \int_0^1 k_{ij}(s,t) f_{ij}(t,u(t))\,dt$$

where

$$f_{ij}(t,u(t)) = f_{ij}(t,u_1(t),u_2(t),\dots,u_n(t)),$$

$i = 1,2,\dots,n$ and $0 \leqslant s \leqslant 1$.

Let $K(s,t) = (k_{ij}(s,t))_{1 \leqslant i,j \leqslant n}$ and

$$F(t,\xi) = (f_{ij}(t,\xi_1,\xi_2,\dots,\xi_n))_{1 \leqslant i,j \leqslant n}$$

be $n \times n$ matrices and let Z be the closed unit square.

APPLICATION 1. Suppose that:

a) $K(s,t)$ is continuous on Z.

b) The elements of the matrix

$$L(t) = \int_0^1 K(s,t)\,ds = \left(\int_0^1 k_{ij}(s,t)\,ds\right)_{1\leqslant i,j\leqslant n}$$

are strictly positive for all t, $0 \leqslant t \leqslant 1$.

Consider the continuous linear functional:

$$\phi: C_o^n[0,1] \longrightarrow \mathbb{R}$$

$$u \longmapsto \sum_{i=1}^{n} \int_0^1 u_i(s)\,ds$$

and the closed affine hyperplane

$$H = \{u \in C_o^n[0,1] \mid \phi_u = 1\}.$$

Let

$$M = \max_{1\leqslant i,j\leqslant n} \max_{Z} \frac{|k_{ij}(s,t)|}{\int_0^1 k_{ij}(s,t)\,ds}$$

and $M_1 \geqslant M$ be any number such that the set $A_1 = H \cap \overline{B}_1$ is non-empty, where $\overline{B}_1 = \{u \in C_o^n[0,1] \mid \|u\| \leqslant M_1\}$.

Assume that:

c) The matrix $F(t,\xi)$ is continuous for $0 \leq t \leq 1$ and $\|\xi\| \leq M_1$.

d) The elements of the matrix $F(t,\xi)$ are non-negative and at least one of them is strictly positive for $0 \leq t \leq 1$ and $\|\xi\| \leq M_1$.

The operator $T: A_1 \longrightarrow C_o^n[0,1]$ defined, for $0 \leq s \leq 1$, as

$$Tu(s) = v(s) = (v_1(s), v_2(s), \ldots, v_n(s))$$

where

$$v_i(s) = \sum_{j=1}^{n} \int_0^1 k_{ij}(s,t) f_{ij}(t,u(t))\,dt$$

is continuous and compact on A_1, because of hypotheses (a) and (c).

Moreover the conditions (i) and (ii) of Theorem 1 are satisfied. In fact, let $u \in A_1$. Then from hypotheses (b) and (d) it follows that:

$$\phi(Tu) = \sum_{i=1}^{n} \int_0^1 \left[\sum_{j=1}^{n} \int_0^1 k_{ij}(s,t) f_{ij}(t,u(t))\,dt \right] ds =$$

$$= \sum_{i,j=1}^{n} \int_0^1 f_{ij}(t,u(t))\,dt \left[\int_0^1 k_{ij}(s,t)\,ds \right] > 0.$$

On the other hand:

$$\|Tu\| = \sum_{i=1}^{n} \max_s \left| \sum_{j=1}^{n} \int_0^1 k_{ij}(s,t) f_{ij}(t,u(t))\,dt \right| \leq$$

$$\leq \sum_{i,j=1}^{n} \max_s \int_0^1 |k_{ij}(s,t)|\, f_{ij}(t,u(t))\,dt =$$

$$= \sum_{i,j=1}^{n} \max_{s} \int_0^1 \frac{k_{ij}(s,t)}{\int_0^1 k_{ij}(s,t)ds} f_{ij}(t,u(t)) \left[\int_0^1 k_{ij}(s,t)ds \right] dt \leqslant$$

$$\leqslant \sum_{i,j=1}^{n} \max_{s} \int_0^1 \left[\max_{t} \frac{|k_{ij}(s,t)|}{\int_0^1 k_{ij}(s,t)ds} \right] \left[\int_0^1 k_{ij}(s,t) f_{ij}(t,u(t))ds \right] dt \leqslant$$

$$\leqslant M \sum_{i,j=1}^{n} \int_0^1 \left[\int_0^1 k_{ij}(,st) f_{ij}(t,u(t))ds \right] dt = M\phi(Tu).$$

By applying Theorem 1 we obtain the eigenfunction $\bar{U}$ satisfying

$$\sum_{i=1}^{n} \int_0^1 \bar{u}_i(s)ds = 1,$$

$\|\bar{u}\| \leqslant M_1$ and $\lambda_o \bar{u} = T\bar{u}$ with

$$\lambda_o = \phi(T\bar{u}) = \sum_{i,j=1}^{n} \int_0^1 \left[\int_0^1 k_{ij}(s,t) f_{ij}(t,u(t))dt \right] ds > 0.$$

APPLICATION 2. We consider the same situation as in application 1 and maintain hypotheses (a) but hypotheses (b) is changed to

(b') The elements of the matrix $K(s_o,t)$ are strictly positive, for some fixed $s_o \in [0,1]$.

Hypotheses (c) and (d) are the same as above but the constant M is now obtained in the following manner:

$$M = \max_{1 \leqslant i,j \leqslant n} \max_{Z} \frac{|k_{ij}(s,t)|}{k_{ij}(s_o,t)} .$$

The continuous linear functional ϕ on $C_o^n[0,1]$ is now given by the following expression:

$$\phi u = \sum_{i=1}^{n} u_i(s_o).$$

In order to verify that conditions (i) and (ii) of Theorem 1 are satisfied, we observe that:

$$\phi(Tu) = \sum_{i=1}^{n} \int_0^1 k_{ij}(s_o,t) f_{ij}(t,u(t))dt$$

is clearly greater than zero.

Furthermore:

$$\|Tu\| = \sum_{i=1}^{n} \max_{s} \left| \sum_{j=1}^{n} \int_0^1 k_{ij}(s,t) f_{ij}(t,u(t))dt \right| \leqslant$$

$$\leqslant \sum_{i,j=1}^{n} \max_{s} \int_0^1 |k_{ij}(s,t)| f_{ij}(t,u(t))dt =$$

$$= \sum_{i,j=1}^{n} \max_{s} \int_0^1 \frac{|k_{ij}(s,t)|}{k_{ij}(s_o,t)} k_{ij}(s_o,t) f_{ij}(t,u(t))dt \leqslant$$

$$\leqslant \sum_{i,j=1}^{n} \max_{s} \int_0^1 \left[\max_t \frac{k_{ij}(s,t)}{k_{ij}(s_o,t)}\right] k_{ij}(s_o,t) f_{ij}(t,u(t))\,dt \leqslant$$

$$\leqslant M \sum_{i,j=1}^{n} \int_0^1 k_{ij}(s_o,t) f_{ij}(t,u(t))\,dt = M\phi(Tu)$$

and again an application of Theorem 1 provides the existence of an eigenfunction $\bar{u}$ which now satisfies

$$\sum_{i=1}^{n} \bar{u}_i(s_o,t) = 1,$$

$\|u\| \leqslant M_1$ and $\lambda_o \bar{u} = T\bar{u}$, with

$$\lambda_o = \phi(T\bar{u}) = \sum_{i,j=1}^{n} \int_0^1 k_{ij}(s_o,t) f_{ij}(t,\bar{u}(t))\,dt.$$

SECTION 3

In trying to weaken the hypotheses on the functional ϕ of Theorem 1 in order to include cases where ϕ is nonlinear we would still like to obtain an eigenvector x_o of the operator T satisfying $\phi x_o = 1$. For this purpose we shall make use of the Leray-Schauder Degree Theory of Compact Operators in Banach spaces and prove:

THEOREM 2. *Let* X *be a Banach space and* $\phi: X \longrightarrow \mathbb{R}$ *a continuous functional such that* $\phi(0) < 1$. *Let* $G = \{x \in X \mid \phi x < 1\}$ *and suppose that* $T: \bar{G} \longrightarrow X$ *is a continuous and compact operator. If the following conditions hold true on* $\bar{G}$:

i) $\phi(Tx) \geqslant 1$

ii) $\exists\, M > 0$ *such that* $\|Tx\| \leqslant M[\phi(Tx)]$

iii) ϕ *is positively subhomogeneous of degree one, that is,*

$$\phi(\lambda x) \leqslant \lambda \phi x,$$

for all $\lambda \geqslant 0$.

Then there exists $\lambda_o > 1$ *and there exists* x_o *with* $\phi x_o = 1$ *such that* $\lambda_o x_o = Tx_o$.

PROOF. We can always find a constant $M_1 \geqslant M$ such that the closed ball $\bar{B}_1 = \{x \in X \mid \|x\| \leqslant M_1\}$ intersects ∂G. Let $U = G \cap B_1$, where B_1 is the interior of the closed ball $\bar{B}_1$. Then U is a bounded open set in X which contains the origin. $\partial U = F_1 \cup F_2$, $F_1 \cap F_2 = \emptyset$, where $F_1 = \partial G \cap \bar{B}_1$ and $F_2 = G \cap \partial B_1$.

Next we show that the homotopy of compact operators:

$$h_\alpha(T) = I - \alpha T: \ \bar{U} \longrightarrow X, \quad 0 \leqslant \alpha \leqslant 1$$

never vanishes on the boundary of U.

This is of course the case if $\alpha = 0$, since $\phi(0) < 1$ and $0 \notin \partial B_1$.

If there exists $x_o \in F_2$ such that $x_o - \alpha_o Tx_o = 0$ for some $\alpha_o \in (0,1]$, we would have $\lambda_o x_o = Tx_o$ with $\lambda_o \geqslant 1$. But, since $x_o \in \partial B_1$, it follows from (ii) and (iii):

$$\lambda_o M_1 = \lambda_o \|x_o\| = \|\lambda_o x_o\| = \|Tx_o\| \leqslant M\phi(Tx_o) =$$

$$= M\phi(\lambda_o x_o) \leqslant \lambda_o M\phi x_o$$

hence $\phi x_o \geqslant M_1/M \geqslant 1$, and $x_o \notin G$ and a fortiori $x_o \notin F_2$.

Now let us assume that the conclusion of the theorem is false, i. e., assume that, for every $\lambda \geqslant 1$, there exists no $x \in \partial G$ such that $\lambda x = Tx$.

Then $\deg(h_\alpha(T), U, 0)$ is invariant with α, $0 \leqslant \alpha \leqslant 1$. Since the only solution of $h_o(T)x = 0$ in U is $x = 0$, we conclude that

$$\deg(h_\alpha(T), U, 0) = \pm 1.$$

Hence, there exists $\bar{x} \in \bar{U}$ such that $\bar{x} = T\bar{x}$ or, because of (i), $\phi\bar{x} = \phi(T\bar{x}) \geqslant 1$ thus implying $\bar{x} \notin G$ or, that $\bar{x} \in \partial G$, which is a contradiction, and the theorem is proved.

Universidade Federal de Pernambuco
Instituto de Matemática
Recife, PE
BRASIL

BIBLIOGRAPHY

[1] J.CRONIN, *Fixed points and topological degree in nonlinear Analysis*, Amer. Math. Soc. Surveys, Nº 11 (1964).

[2] S.KARLIN and L.NIRENBERG, *On a theorem of Nowosad*, Jour. Math. Anal. & Appl., vol. 17, Nº 1 (1967).

[3] M.KRASNOSELSKII, *Topological Methods in the Theory of Nonlinear Integral Equations*, Moscow (1956), Eng. Transl. Mac Millan, New York (1964).

[4] R.RAMALHO, *On the existence theorems for nonlinear integral equations*, Ph. D. Thesis, Courant Institute of Mathematical Sciences, New York University (1968).

Vol. 215: P. Antonelli, D. Burghelea and P. J. Kahn, The Concordance-Homotopy Groups of Geometric Automorphism Groups. X, 140 pages. 1971. DM 16,-

Vol. 216: H. Maaß, Siegel's Modular Forms and Dirichlet Series. VII, 328 pages. 1971. DM 20,-

Vol. 217: T. J. Jech, Lectures in Set Theory with Particular Emphasis on the Method of Forcing. V, 137 pages. 1971. DM 16,-

Vol. 218: C. P. Schnorr, Zufälligkeit und Wahrscheinlichkeit. IV, 212 Seiten. 1971. DM 20,-

Vol. 219: N. L. Alling and N. Greenleaf, Foundations of the Theory of Klein Surfaces. IX, 117 pages. 1971. DM 16,-

Vol. 220: W. A. Coppel, Disconjugacy. V, 148 pages. 1971. DM 16,-

Vol. 221: P. Gabriel und F. Ulmer, Lokal präsentierbare Kategorien. V, 200 Seiten. 1971. DM 18,-

Vol. 222: C. Meghea, Compactification des Espaces Harmoniques. III, 108 pages. 1971. DM 16,-

Vol. 223: U. Felgner, Models of ZF-Set Theory. VI, 173 pages. 1971. DM 16,-

Vol. 224: Revètements Etales et Groupe Fondamental. (SGA 1). Dirigé par A. Grothendieck XXII, 447 pages. 1971. DM 30,-

Vol. 225: Théorie des Intersections et Théorème de Riemann-Roch. (SGA 6). Dirigé par P. Berthelot, A. Grothendieck et L. Illusie. XII, 700 pages. 1971. DM 40,-

Vol. 226: Seminar on Potential Theory, II. Edited by H. Bauer. IV, 170 pages. 1971. DM 18,-

Vol. 227: H. L. Montgomery, Topics in Multiplicative Number Theory. IX, 178 pages. 1971. DM 18,-

Vol. 228: Conference on Applications of Numerical Analysis. Edited by J. Ll. Morris. X, 358 pages. 1971. DM 26,-

Vol. 229: J. Väisälä, Lectures on n-Dimensional Quasiconformal Mappings. XIV, 144 pages. 1971. DM 16,-

Vol. 230: L. Waelbroeck, Topological Vector Spaces and Algebras. VII, 158 pages. 1971. DM 16,-

Vol. 231: H. Reiter, L^1-Algebras and Segal Algebras. XI, 113 pages. 1971. DM 16,-

Vol. 232: T. H. Ganelius, Tauberian Remainder Theorems. VI, 75 pages. 1971. DM 16,-

Vol. 233: C. P. Tsokos and W. J. Padgett. Random Integral Equations with Applications to stochastic Systems. VII, 174 pages. 1971. DM 18,-

Vol. 234: A. Andreotti and W. Stoll. Analytic and Algebraic Dependence of Meromorphic Functions. III, 390 pages. 1971. DM 26,-

Vol. 235: Global Differentiable Dynamics. Edited by O. Hájek, A. J. Lohwater, and R. McCann. X, 140 pages. 1971. DM 16,-

Vol. 236: M. Barr, P. A. Grillet, and D. H. van Osdol. Exact Categories and Categories of Sheaves. VII, 239 pages. 1971. DM 20,-

Vol. 237: B. Stenström, Rings and Modules of Quotients. VII, 136 pages. 1971. DM 16,-

Vol. 238: Der kanonische Modul eines Cohen-Macaulay-Rings. Herausgegeben von Jürgen Herzog und Ernst Kunz. VI, 103 Seiten. 1971. DM 16,-

Vol. 239: L. Illusie, Complexe Cotangent et Déformations I. XV, 355 pages. 1971. DM 26,-

Vol. 240: A. Kerber, Representations of Permutation Groups I. VII, 192 pages. 1971. DM 18,-

Vol. 241: S. Kaneyuki, Homogeneous Bounded Domains and Siegel Domains. V, 89 pages. 1971. DM 16,-

Vol. 242: R. R. Coifman et G. Weiss, Analyse Harmonique Non-Commutative sur Certains Espaces. V, 160 pages. 1971. DM 16,-

Vol. 243: Japan-United States Seminar on Ordinary Differential and Functional Equations. Edited by M. Urabe. VIII, 332 pages. 1971. DM 26,-

Vol. 244: Séminaire Bourbaki - vol. 1970/71. Exposés 382-399. IV, 356 pages. 1971. DM 26,-

Vol. 245: D. E. Cohen, Groups of Cohomological Dimension One. V, 99 pages. 1972. DM 16,-

Vol. 246: Lectures on Rings and Modules. Tulane University Ring and Operator Theory Year, 1970-1971. Volume I. X, 661 pages. 1972. DM 40,-

Vol. 247: Lectures on Operator Algebras. Tulane University Ring and Operator Theory Year, 1970-1971. Volume II. XI, 786 pages. 1972. DM 40,-

Vol. 248: Lectures on the Applications of Sheaves to Ring Theory. Tulane University Ring and Operator Theory Year, 1970-1971. Volume III. VIII, 315 pages. 1971. DM 26,-

Vol. 249: Symposium on Algebraic Topology. Edited by P. J. Hilton. VII, 111 pages. 1971. DM 16,-

Vol. 250: B. Jónsson, Topics in Universal Algebra. VI, 220 pages. 1972. DM 20,-

Vol. 251: The Theory of Arithmetic Functions. Edited by A. A. Gioia and D. L. Goldsmith VI, 287 pages. 1972. DM 24,-

Vol. 252: D. A. Stone, Stratified Polyhedra. IX, 193 pages. 1972. DM 18,-

Vol. 253: V. Komkov, Optimal Control Theory for the Damping of Vibrations of Simple Elastic Systems. V, 240 pages. 1972. DM 20,-

Vol. 254: C. U. Jensen, Les Foncteurs Dérivés de $\varprojlim$ et leurs Applications en Théorie des Modules. V, 103 pages. 1972. DM 16,-

Vol. 255: Conference in Mathematical Logic - London '70. Edited by W. Hodges. VIII, 351 pages. 1972. DM 26,-

Vol. 256: C. A. Berenstein and M. A. Dostal, Analytically Uniform Spaces and their Applications to Convolution Equations. VII, 130 pages. 1972. DM 16,-

Vol. 257: R. B. Holmes, A Course on Optimization and Best Approximation. VIII, 233 pages. 1972. DM 20,-

Vol. 258: Séminaire de Probabilités VI. Edited by P. A. Meyer. VI, 253 pages. 1972. DM 22,-

Vol. 259: N. Moulis, Structures de Fredholm sur les Variétés Hilbertiennes. V, 123 pages. 1972. DM 16,-

Vol. 260: R. Godement and H. Jacquet, Zeta Functions of Simple Algebras. IX, 188 pages. 1972. DM 18,-

Vol. 261: A. Guichardet, Symmetric Hilbert Spaces and Related Topics. V, 197 pages. 1972. DM 18,-

Vol. 262: H. G. Zimmer, Computational Problems, Methods, and Results in Algebraic Number Theory. V, 103 pages. 1972. DM 16,-

Vol. 263: T. Parthasarathy, Selection Theorems and their Applications. VII, 101 pages. 1972. DM 16,-

Vol. 264: W. Messing, The Crystals Associated to Barsotti-Tate Groups: With Applications to Abelian Schemes. III, 190 pages. 1972. DM 18,-

Vol. 265: N. Saavedra Rivano, Catégories Tannakiennes. II, 418 pages. 1972. DM 26,-

Vol. 266: Conference on Harmonic Analysis. Edited by D. Gulick and R. L. Lipsman. VI, 323 pages. 1972. DM 24,-

Vol. 267: Numerische Lösung nichtlinearer partieller Differential- und Integro-Differentialgleichungen. Herausgegeben von R. Ansorge und W. Törnig, VI, 339 Seiten. 1972. DM 26,-

Vol. 268: C. G. Simader, On Dirichlet's Boundary Value Problem. IV, 238 pages. 1972. DM 20,-

Vol. 269: Théorie des Topos et Cohomologie Etale des Schémas. (SGA 4). Dirigé par M. Artin, A. Grothendieck et J. L. Verdier. XIX, 525 pages. 1972. DM 50,-

Vol. 270: Théorie des Topos et Cohomologie Etale des Schémas. Tome 2. (SGA 4). Dirigé par M. Artin, A. Grothendieck et J. L. Verdier. V, 418 pages. 1972. DM 50,-

Vol. 271: J. P. May, The Geometry of Iterated Loop Spaces. IX, 175 pages. 1972. DM 18,-

Vol. 272: K. R. Parthasarathy and K. Schmidt, Positive Definite Kernels, Continuous Tensor Products, and Central Limit Theorems of Probability Theory. VI, 107 pages. 1972. DM 16,-

Vol. 273: U. Seip, Kompakt erzeugte Vektorräume und Analysis. IX, 119 Seiten. 1972. DM 16,-

Vol. 274: Toposes, Algebraic Geometry and Logic. Edited by. F. W. Lawvere. VI, 189 pages. 1972. DM 18,-

Vol. 275: Séminaire Pierre Lelong (Analyse) Année 1970-1971. VI, 181 pages. 1972. DM 18,-

Vol. 276: A. Borel, Représentations de Groupes Localement Compacts. V, 98 pages. 1972. DM 16,-

Vol. 277: Séminaire Banach. Edité par C. Houzel. VII, 229 pages. 1972. DM 20,-

Vol. 278: H. Jacquet, Automorphic Forms on GL(2). Part II. XIII, 142 pages. 1972. DM 16,–

Vol. 279: R. Bott, S. Gitler and I. M. James, Lectures on Algebraic and Differential Topology. V, 174 pages. 1972. DM 18,–

Vol. 280: Conference on the Theory of Ordinary and Partial Differential Equations. Edited by W. N. Everitt and B. D. Sleeman. XV, 367 pages. 1972. DM 26,–

Vol. 281: Coherence in Categories. Edited by S. Mac Lane. VII, 235 pages. 1972. DM 20,–

Vol. 282: W. Klingenberg und P. Flaschel, Riemannsche Hilbertmannigfaltigkeiten. Periodische Geodätische. VII, 211 Seiten. 1972. DM 20,–

Vol. 283: L. Illusie, Complexe Cotangent et Déformations II. VII, 304 pages. 1972. DM 24,–

Vol. 284: P. A. Meyer, Martingales and Stochastic Integrals I. VI, 89 pages. 1972. DM 16,–

Vol. 285: P. de la Harpe, Classical Banach-Lie Algebras and Banach-Lie Groups of Operators in Hilbert Space. III, 160 pages. 1972. DM 16,–

Vol. 286: S. Murakami, On Automorphisms of Siegel Domains. V, 95 pages. 1972. DM 16,–

Vol. 287: Hyperfunctions and Pseudo-Differential Equations. Edited by H. Komatsu. VII, 529 pages. 1973. DM 36,–

Vol. 288: Groupes de Monodromie en Géométrie Algébrique. (SGA 7 I). Dirigé par A. Grothendieck. IX, 523 pages. 1972. DM 50,–

Vol. 289: B. Fuglede, Finely Harmonic Functions. III, 188. 1972. DM 18,–

Vol. 290: D. B. Zagier, Equivariant Pontrjagin Classes and Applications to Orbit Spaces. IX, 130 pages. 1972. DM 16,–

Vol. 291: P. Orlik, Seifert Manifolds. VIII, 155 pages. 1972. DM 16,–

Vol. 292: W. D. Wallis, A. P. Street and J. S. Wallis, Combinatorics: Room Squares, Sum-Free Sets, Hadamard Matrices. V, 508 pages. 1972. DM 50,–

Vol. 293: R. A. DeVore, The Approximation of Continuous Functions by Positive Linear Operators. VIII, 289 pages. 1972. DM 24,–

Vol. 294: Stability of Stochastic Dynamical Systems. Edited by R. F. Curtain. IX, 332 pages. 1972. DM 26,–

Vol. 295: C. Dellacherie, Ensembles Analytiques, Capacités, Mesures de Hausdorff. XII, 123 pages. 1972. DM 16,–

Vol. 296: Probability and Information Theory II. Edited by M. Behara, K. Krickeberg and J. Wolfowitz. V, 223 pages. 1973. DM 20,–

Vol. 297: J. Garnett, Analytic Capacity and Measure. IV, 138 pages. 1972. DM 16,–

Vol. 298: Proceedings of the Second Conference on Compact Transformation Groups. Part 1. XIII, 453 pages. 1972. DM 32,–

Vol. 299: Proceedings of the Second Conference on Compact Transformation Groups. Part 2. XIV, 327 pages. 1972. DM 26,–

Vol. 300: P. Eymard, Moyennes Invariantes et Représentations Unitaires. II. 113 pages. 1972. DM 16,–

Vol. 301: F. Pittnauer, Vorlesungen über asymptotische Reihen. VI, 186 Seiten. 1972. DM 18,–

Vol. 302: M. Demazure, Lectures on p-Divisible Groups. V, 98 pages. 1972. DM 16,–

Vol. 303: Graph Theory and Applications. Edited by Y. Alavi, D. R. Lick and A. T. White. IX, 329 pages. 1972. DM 26,–

Vol. 304: A. K. Bousfield and D. M. Kan, Homotopy Limits, Completions and Localizations. V, 348 pages. 1972. DM 26,–

Vol. 305: Théorie des Topos et Cohomologie Etale des Schémas. Tome 3. (SGA 4). Dirigé par M. Artin, A. Grothendieck et J. L. Verdier. VI, 640 pages. 1973. DM 50,–

Vol. 306: H. Luckhardt, Extensional Gödel Functional Interpretation. VI, 161 pages. 1973. DM 18,–

Vol. 307: J. L. Bretagnolle, S. D. Chatterji et P.-A. Meyer, Ecole d'été de Probabilités: Processus Stochastiques. VI, 198 pages. 1973. DM 20,–

Vol. 308: D. Knutson, λ-Rings and the Representation Theory of the Symmetric Group. IV, 203 pages. 1973. DM 20,–

Vol. 309: D. H. Sattinger, Topics in Stability and Bifurcation Theory. VI, 190 pages. 1973. DM 18,–

Vol. 310: B. Iversen, Generic Local Structure of the Morphisms in Commutative Algebra. IV, 108 pages. 1973. DM 16,–

Vol. 311: Conference on Commutative Algebra. Edited by J. W. Brewer and E. A. Rutter. VII, 251 pages. 1973. DM 22,–

Vol. 312: Symposium on Ordinary Differential Equations. Edited by W. A. Harris, Jr. and Y. Sibuya. VIII, 204 pages. 1973. DM 22,–

Vol. 313: K. Jörgens and J. Weidmann, Spectral Properties of Hamiltonian Operators. III, 140 pages. 1973. DM 16,–

Vol. 314: M. Deuring, Lectures on the Theory of Algebraic Functions of One Variable. VI, 151 pages. 1973. DM 16,–

Vol. 315: K. Bichteler, Integration Theory (with Special Attention to Vector Measures). VI, 357 pages. 1973. DM 26,–

Vol. 316: Symposium on Non-Well-Posed Problems and Logarithmic Convexity. Edited by R. J. Knops. V, 176 pages. 1973. DM 18,–

Vol. 317: Séminaire Bourbaki – vol. 1971/72. Exposés 400–417. IV, 361 pages. 1973. DM 26,–

Vol. 318: Recent Advances in Topological Dynamics. Edited by A. Beck, VIII, 285 pages. 1973. DM 24,–

Vol. 319: Conference on Group Theory. Edited by R. W. Gatterdam and K. W. Weston. V, 188 pages. 1973. DM 18,–

Vol. 320: Modular Functions of One Variable I. Edited by W. Kuyk. V, 195 pages. 1973. DM 18,–

Vol. 321: Séminaire de Probabilités VII. Edité par P. A. Meyer. VI, 322 pages. 1973. DM 26,–

Vol. 322: Nonlinear Problems in the Physical Sciences and Biology. Edited by I. Stakgold, D. D. Joseph and D. H. Sattinger. VIII, 357 pages. 1973. DM 26,–

Vol. 323: J. L. Lions, Perturbations Singulières dans les Problèmes aux Limites et en Contrôle Optimal. XII, 645 pages. 1973. DM 42,–

Vol. 324: K. Kreith, Oscillation Theory. VI, 109 pages. 1973. DM 16,–

Vol. 325: Ch.-Ch. Chou, La Transformation de Fourier Complexe et L'Equation de Convolution. IX, 137 pages. 1973. DM 16,–

Vol. 326: A. Robert, Elliptic Curves. VIII, 264 pages. 1973. DM 22,–

Vol. 327: E. Matlis, 1-Dimensional Cohen-Macaulay Rings. XII, 157 pages. 1973. DM 18,–

Vol. 328: J. R. Büchi and D. Siefkes, The Monadic Second Order Theory of All Countable Ordinals. VI, 217 pages. 1973. DM 20,–

Vol. 329: W. Trebels, Multipliers for (C, α)-Bounded Fourier Expansions in Banach Spaces and Approximation Theory. VII, 103 pages. 1973. DM 16,–

Vol. 330: Proceedings of the Second Japan-USSR Symposium on Probability Theory. Edited by G. Maruyama and Yu. V. Prokhorov. VI, 550 pages. 1973. DM 36,–

Vol. 331: Summer School on Topological Vector Spaces. Edited by L. Waelbroeck. VI, 226 pages. 1973. DM 20,–

Vol. 332: Séminaire Pierre Lelong (Analyse) Année 1971-1972. V, 131 pages. 1973. DM 16,–

Vol. 333: Numerische, insbesondere approximationstheoretische Behandlung von Funktionalgleichungen. Herausgegeben von R. Ansorge und W. Törnig. VI, 296 Seiten. 1973. DM 24,–

Vol. 334: F. Schweiger, The Metrical Theory of Jacobi-Perron Algorithm. V, 111 pages. 1973. DM 16,–

Vol. 335: H. Huck, R. Roitzsch, U. Simon, W. Vortisch, R. Walden, B. Wegner und W. Wendland, Beweismethoden der Differentialgeometrie im Großen. IX, 159 Seiten. 1973. DM 18,–

Vol. 336: L'Analyse Harmonique dans le Domaine Complexe. Edité par E. J. Akutowicz. VIII, 169 pages. 1973. DM 18,–

Vol. 337: Cambridge Summer School in Mathematical Logic. Edited by A. R. D. Mathias and H. Rogers. IX, 660 pages. 1973. DM 42,–

Vol. 338: J. Lindenstrauss and L. Tzafriri, Classical Banach Spaces. IX, 243 pages. 1973. DM 22,–

Vol. 339: G. Kempf, F. Knudsen, D. Mumford and B. Saint-Donat, Toroidal Embeddings I. VIII, 209 pages. 1973. DM 20,–

Vol. 340: Groupes de Monodromie en Géométrie Algébrique. (SGA 7 II). Par P. Deligne et N. Katz. X, 438 pages. 1973. DM 40,–

Vol. 341: Algebraic K-Theory I, Higher K-Theories. Edited by H. Bass. XV, 335 pages. 1973. DM 26,–

Vol. 342: Algebraic K-Theory II, "Classical" Algebraic K-Theory, and Connections with Arithmetic. Edited by H. Bass. XV, 527 pages. 1973. DM 36,–